Springer Series in Translational Stroke Research

Series Editor
John Zhang
Loma Linda, CA, USA

More information about this series at http://www.springer.com/series/10064

Jun Chen • Jian Wang • Ling Wei
John H. Zhang
Editors

Therapeutic Intranasal Delivery for Stroke and Neurological Disorders

Editors
Jun Chen
Department of Neurology
University of Pittsburgh
Pittsburgh, PA, USA

Ling Wei
Departments of Anesthesiology
and Neurology
Emory University School of Medicine
Atlanta, GA, USA

Jian Wang
Department of Anesthesiology
and Critical Care Medicine
The Johns Hopkins University School
of Medicine
Baltimore, MD, USA

John H. Zhang
Department of Physiology
Loma Linda University School of Medicine
Loma Linda, CA, USA

ISSN 2363-958X	ISSN 2363-9598	(electronic)
Springer Series in Translational Stroke Research
ISBN 978-3-030-16713-4	ISBN 978-3-030-16715-8	(eBook)
https://doi.org/10.1007/978-3-030-16715-8

© Springer Nature Switzerland AG 2019
This work is subject to copyright. All rights are reserved by the Publisher, whether the whole or part of the material is concerned, specifically the rights of translation, reprinting, reuse of illustrations, recitation, broadcasting, reproduction on microfilms or in any other physical way, and transmission or information storage and retrieval, electronic adaptation, computer software, or by similar or dissimilar methodology now known or hereafter developed.
The use of general descriptive names, registered names, trademarks, service marks, etc. in this publication does not imply, even in the absence of a specific statement, that such names are exempt from the relevant protective laws and regulations and therefore free for general use.
The publisher, the authors, and the editors are safe to assume that the advice and information in this book are believed to be true and accurate at the date of publication. Neither the publisher nor the authors or the editors give a warranty, express or implied, with respect to the material contained herein or for any errors or omissions that may have been made. The publisher remains neutral with regard to jurisdictional claims in published maps and institutional affiliations.

This Springer imprint is published by the registered company Springer Nature Switzerland AG
The registered company address is: Gewerbestrasse 11, 6330 Cham, Switzerland

Contents

1 **Transnasal Induction of Therapeutic Hypothermia for Neuroprotection** 1
   Raghuram Chava and Harikrishna Tandri

2 **Hypoxia-Primed Stem Cell Transplantation in Stroke** 9
   Zheng Zachory Wei, James Ya Zhang, and Ling Wei

3 **Therapeutic Potential of Intranasal Drug Delivery in Preclinical Studies of Ischemic Stroke and Intracerebral Hemorrhage** 27
   Qian Li, Claire F. Levine, and Jian Wang

4 **Intranasal Drug Delivery After Intracerebral Hemorrhage** 43
   Jing Chen-Roetling and Raymond F. Regan

5 **Intranasal Treatment in Subarachnoid Hemorrhage** 57
   Basak Caner

6 **Intranasal Delivery of Therapeutic Peptides for Treatment of Ischemic Brain Injury** 65
   Tingting Huang, Amanda Smith, Jun Chen, and Peiying Li

7 **Intranasal Delivering Method in the Treatment of Ischemic Stroke** 75
   Chunhua Chen, Mengqin Zhang, Yejun Wu, Changman Zhou, and Renyu Liu

8 **Intranasal Delivery of Drugs for Ischemic Stroke Treatment: Targeting IL-17A** 91
   Yun Lin, Jiancheng Zhang, and Jian Wang

9 **Intranasal tPA Application for Axonal Remodeling in Rodent Stroke and Traumatic Brain Injury Models** 101
   Zhongwu Liu, Ye Xiong, and Michael Chopp

10	**Therapeutic Intranasal Delivery for Alzheimer's Disease**.......... Xinxin Wang and Fangxia Guan	117
11	**Intranasal Medication Delivery in Children for Brain Disorders** ... Gang Zhang, Myles R. McCrary, and Ling Wei	135

Index.. 149

Chapter 1
Transnasal Induction of Therapeutic Hypothermia for Neuroprotection

Raghuram Chava and Harikrishna Tandri

Abstract Effective control of core body temperature and producing hypothermia is the standard of care for comatose patients with cardiac arrest and also in neurogenic fevers. Nasopharyngeal space has been a region of great interest to induce therapeutic hypothermia for a long time. This is primarily due to the favorable location of the nasal heat exchanger directly beneath the brain, the main target for hypothermia. This chapter focuses on achieving therapeutic hypothermia of the brain and core body temperature by using transnasal dry air.

Keywords Therapeutic hypothermia · Selective cerebral cooling · Transnasal cooling

Nasopharyngeal space has been a region of great interest to induce therapeutic hypothermia for a long time. This is primarily due to the favorable location of the nasal heat exchanger directly beneath the brain, the main target for hypothermia. Nasal heat exchanger is a highly evolved system and has undergone significant evolutionary changes. The architecture of the nasal passages, the size of the turbinates (primary site for heat exchange) vary significantly among species and within the same species based on the environment. In human beings, people inhabiting the polar zones often have long nasal passages and complex nasal turbinate morphology that enables optimal conditioning of the inspired air, a prerequisite for the health of the lower respiratory tract. People in the tropical climates typically have shorter and wider nasal passages as conditioning is less of a burden in hot humid climates. The nasal mucosa is also highly evolved and is supplied by a rich plexus of venous channels which form submucosal sinusoidal spaces that are optimal for heat exchange [1]. Both the internal and external carotid arteries supply the nasal cavity. The anterior and posterior ethmoid arteries, both branches of the internal carotid artery system supply the upper

---

R. Chava · H. Tandri (✉)
Center for Cardiac Innovations, Division of Cardiology, Johns Hopkins University School of Medicine, Baltimore, MD, USA
e-mail: htandril@jhmi.edu

nasal septum and nasal sidewalls. The superior labial branch of the facial artery supplies the front part of the nose. The sphenopalatine artery, a branch of the external carotid system supplies most of the back of the nasal cavity.

The nasal turbinates have high capacity to engorge and increase significantly in surface area to promote heat exchange. Nasal mucosa also has a rich submucosal goblet cell layer that secretes nasal mucus. Evaporation of nasal mucus is the primary mechanism that drives the heat exchange function of the nose. Nasal mucosal glands are capable of secreting up to 1 mL/min of mucus fluid. An average person breathing 6–7 L/min consumes evaporates up to 600 mL of water in the nasal cavity. Ninety percent of this evaporated water condenses back on to the nasal mucosa during expiration as the expired air is colder than the mucosal surface. Thus, the nasal mucosa conserves water loss. Mucus production is significantly increased on exposure to irritants, dry air, increased nasal osmolarity and cold weather, thus increasing humidification in such settings. The cranial sinuses such as the maxillary and ethmoid sinus also participate in heat exchange. The internal carotid artery enters the spend air sinus and traverses the venous sinusoidal space in this region. In animals with "carotid rate" an intricate network of connections between the venous sinuses and the carotid artery, this arrangement favors significant exchange of heat from the cerebral arteries to the venous sinuses resulting in selective cerebral cooling. This has not been demonstrated in mammals without carotid rate, which includes humans. However, it is undeniable that the proximity of the nasal heat exchanger does favor local conductive cooling to critical deep brain structures and makes it enticing to exploit this for inducing hypothermia.

Several investigators have explored the possibility of inducing hypothermia through the nasopharynx. Most of the methods relied on instilling cold fluids in to the nasal cavities to cool the nasal mucosa which will in turn cool the paranasal space. Covaciu et al. used cold saline irrigated through a series of thin walled balloons deployed in the nasal cavities of awake volunteers [2]. Balloons were inserted under local anesthesia. Saline at 4 °C was circulated through the balloons. Volunteers underwent MR thermography to assess changes in brain temperature. Over a period of 2 h, up to 2 °C reduction in brain temperature was noted. Except for nasal erythema and discomfort there were no adverse events. No changes were noted in core rectal temperatures in the volunteers suggesting that this was selective cerebral cooling. This was the first study to use MR thermography to assess regional brain temperatures during nasopharyngeal cooling.

Andrews and Harris et al. [3] used flow of ambient air twice the patients minute ventilation in brain injured patients and noted a small decline in brain temperature. The authors used nitric oxide (NO) to promote vasodilation of the nasal mucosa. The temperature of the air was 23 °C with a relative humidity of ~30–35%. Brain temperature was measured using a Camino catheter placed in the prefrontal cortex. Core temperature was measured using an esophageal probe. The intervention lasted for 5 min and the mean airflow was 17.7 LPM. A small but clinically insignificant of 0.2 °C decline in cerebral temperature was note by the investigators which led them to conclude that selective cerebral cooling does not occur. It will be clear in the later part of this report as to why this study failed to show changes in brain temperature.

1.1 Rhinochill

The most promising technology to date that has been studied well in acute brain injury is "Rhinochill" which uses a coolant spray delivered to the nasal cavity that is evaporated by high flow of oxygen [4]. The Rhinochill device (Fig. 1.1) consists of a control unit, a source of compressed oxygen and the coolant bottle which is a proprietary perflurocarbon (PFC). A pair of elongated nasal tubes that are inserted in to the nostrils delivers PFC directly to the turbinates. High flow of oxygen delivered through the tubing set evaporates the PFC that is sprayed in to the nostrils. Rhinochill device has been tested in cardiac arrest [4–7] and in the neurocritical care [8] (NCCU) settings and shows promise in reducing brain temperatures selectively at least in the NCCU cohort. The coolant, a perflurocarbon, has a very low boiling point and belongs to a class of chemicals that are biologically non-reactive. Latent heat of vaporization of the coolant is approximately 85 kJ/kg which is the dominant mechanism behind the heat loss by evaporative cooling in the nasopharynx. In a randomized large out of hospital cardiac arrest study in 200 patients, Rhinochill showed the ability to initiate cooling during resuscitation prior to return of spontaneous circulation, and showed small but significant lowering of tympanic and core temperature in the cooled subjects. No significant adverse effects were observed except for a "white nose" due to excessive cooling which returned to baseline with discontinuation of cooling. In a study in the neurocritical care setting where brain temperature monitoring was feasible, Rhinochill demonstrated a gradient of cooling with maximal cooling occurring in the brain tissue followed by the core body temperature measured in the bladder with a difference of up to 0.5 °C. Further there was also a favorable reduction in the intracranial pressure in brain cooled patients.

Despite positive encouraging studies, and the feasibility studies some concerns with the use of PFC remain a hindrance to its widespread use. PFC chemicals are among the least acutely toxic compounds known, although they are known to be

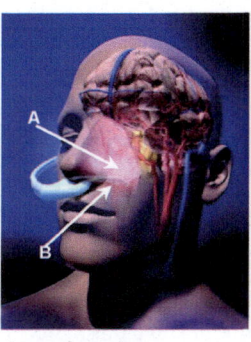
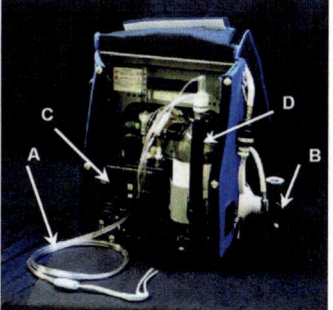

Fig. 1.1 Rhinochill device

A: coolant spray
B: nasal catheter

A: nasal catheter
B: oxygen tank
C: control unit
D: coolant bottle

potent immunomodulators and inhibit white blood cell chemotaxis at femtogram concentrations [9–11]. PFC has also been linked to carcinogenicity in smaller studies and the EPA considers PFC as a toxic substance with detrimental biologic effects [12]. Other than the biologic effects, the PFC has resulted in inadvertent freezing injury of the nasal mucosa which does not completely resolve in all patients with cessation of cooling [6]. Other complications include direct injury of the nasal mucosa by the long nasal cannula and pneumocephalus as the cannulas are inserted by non-specialized medical personnel [13].

Finally, the cost of PFC is prohibitive for use in the ambulance, which is generally a low resource setting. Duration of nasopharyngeal cooling in their clinical trial has been 60 min, and the amount of refrigerant per-patient used was approximately 3.5 L [5]. This would amount to at least $2000 for the consumables alone per hour of use which poses significant strain on the EMS. Rhinochill device is currently not approved for use in the United States.

1.2 Transnasal Evaporative Cooling Using Dry Air: (CoolStat Device)

Evaporative heat loss is a very well understood thermodynamic phenomenon and the human body is uniquely designed to exploit this mechanism for thermoregulation. Evaporation of nasal mucus is fundamental to the humidification of the inspired air by the nasal turbinates and this process can be harnessed to promote heat loss. The latent heat of evaporation of water is 2257 kJ/kg which is among the highest in all known fluids. Einer-Jensen et al. [14] were the first to demonstrate cerebral cooling in intubated pigs by flushing the nostrils with air at high flow rates. However, the authors concluded that this was due to the presence unique vascular plexus in the pigs ("rete mirabile") and the underlying determinants of the cooling were not appreciated. We performed similar experiments in intubated porcine animals and showed that the primary mechanism behind heat loss is the evaporation of nasal mucosal water [15]. The determinants of water evaporation by a dry gas such as the air flow rate and the air humidity govern the cooling response. We not only showed that uniform cerebral cooling occurs (Fig. 1.2), but also systemic cooling ensues with continued exposure to dry air. Humidification of the inspired air abolished the cooling response thus proving that water evaporation is key to this process. Rates of cerebral and systemic cooling in pigs using dry air are comparable to the published perflurocarbon based cooling results. Based on the success of cooling in animals, we performed a human proof of concept study using a prototype device (CoolStat, Fig. 1.3) in 16 intubated subjects in the peri-operative setting (unpublished data) which showed 0.7 °C of reduction in core esophageal temperature over a 1-h period which is comparable to the human data published using the PFC based cooling device. A clinical trial in patients is currently underway to validate this method in the setting of fevers in the neurocritical care unit using this novel device.

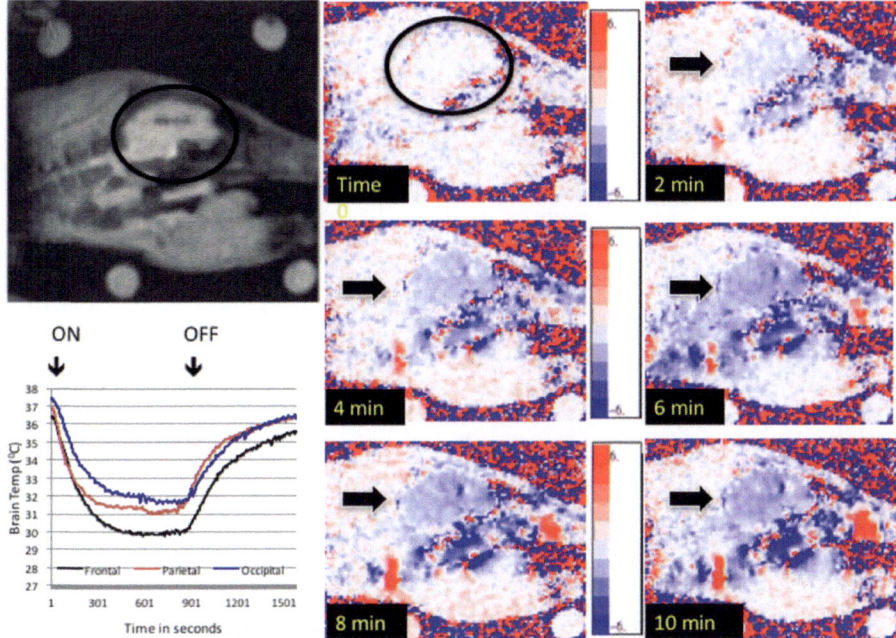

Fig. 1.2 Uniform cooling of a porcine brain using high flow dry air. Shown in the panel in the left top corner is an MRI image of a pig brain in the black circle. On the right is shown the same image over 10 min of dry air flow at 60 LPM through the nostrils. The brain gradually and uniformly cools (black arrows) and note the cooling of the oral nasal cavity. In the bottom left corner is direct measurements of temperature via thermocouples placed in frontal, parietal and temporal cortex showing uniform and rapid brain cooling

Fig. 1.3 CoolStat Device

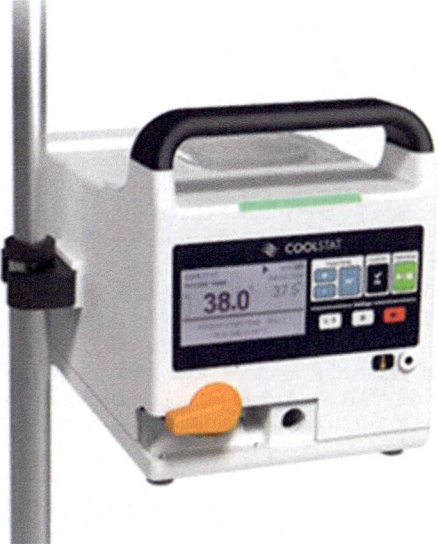

In conclusion, transnasal cooling has gained significant interest and acceptance mainly due to emergence of PFC based cooling in the recent years. This has the potential to induce hypothermia in an out of hospital setting that is particularly attractive in conditions such as cardiac arrest and traumatic brain injury where early intervention has shown significant neuroprotection at least in animal studies. Large scale clinical trials are needed to evaluate the impact of these new devices on survival and neuroprotection in threatened brain injury.

Acknowledgements This work was supported by a NIH SBIR grant to Harikrishna Tandri (1 R44 HL108542-01A1).

Disclosures Dr. Tandri holds patents for transnasal cooling and invented the transnasal dry air cooling device (CoolStat).

References

1. Geurkink N. Nasal anatomy, physiology, and function. J Allergy Clin Immunol. 1983;72(2):123–8.
2. Covaciu L, Allers M, Enblad P, et al. Intranasal selective brain cooling in pigs. Resuscitation. 2008;76:83–8.
3. Andrews PJD, Harris B, Murray GD. A randomised cross-over trial of the effects of airflow through the upper respiratory tract of intubated, brain injured patients on brain temperature and selective brain cooling. Br J Anaesth. 2005;94:330–5.
4. Buscha H, Eichwedeb F, Födisch M, Taccone FS, et al. Safety and feasibility of nasopharyngeal evaporative cooling in the emergency department setting in survivors of cardiac arrest. Resuscitation. 2010;81:943–9.
5. Castren M, Nordberg P, Svensson L, et al. Intra-arrest transnasal evaporative cooling a randomized, prehospital, multicenter study (PRINCE: pre-ROSC intranasal cooling effectiveness). Circulation. 2010;122:729–36.
6. Bollera M, Lampea JW, Katza JM, Barbutc D, Beckera LB. Feasibility of intra-arrest hypothermia induction: a novel nasopharyngeal approach achieves preferential brain cooling. Resuscitation. 2010;81:1025–30.
7. Guan J, Barbut D, Wang H, Li Y, Tsai M, Sun S, Inderbitzen B, Weil MH, Tang W. Rapid induction of head cooling by the intranasal route during cardiopulmonary resuscitation improves survival and neurological outcomes. Crit Care Med. 2008;36(suppl):S428–33.
8. Abou-Chebl A, Sung G, Barbut D, Torbey M. Local brain temperature reduction through intranasal cooling with the RhinoChill device: preliminary safety data in brain-injured patients. Stroke. 2011;42(8):2164–9.
9. Edwards C, Lowe KC, Röhlke W, Geister U, Reuter P, Meinert H. Effects of a novel perfluorocarbon emulsion on neutrophil chemiluminescence in human whole blood in vitro. Artif Cells Blood Substit Immobil Biotechnol. 1997;25:255–60.
10. Flaim S. Pharmacokinetics and side effects of perfluorocarbon-based blood substitutes. Artif Cells Blood Substit Immobil Biotechnol. 1994;22:1043–54.
11. Bucala R, Kawakami M, Cerami A. Cytotoxicity of a perfluorocarbon blood substitute to macrophages in vitro. Science. 1983;27:965–7.
12. OECD/UNEP Global PFC Group. Synthesis paper on per- and polyfluorinated chemicals (PFCs). Paris: Environment, Health and Safety, Environment Directorate, OECD; 2013.

13. Harris S, Bansbach J, Dietrich I, Kalbhenn J, Schmutz A. RhinoChill(®)-more than an "ice-cream headache (1)" serious adverse event related to transnasal evaporative cooling. Resuscitation. 2016;103:e5–6.
14. Einer-Jensen N, Khorooshi MH. Cooling of the brain through oxygen flushing of the nasal cavities in intubated rats: an alternative model for treatment of brain injury. Exp Brain Res. 2000;130:244–7.
15. Chava R, Zviman M, Raghavan MS, Halperin H, et al. Rapid induction of therapeutic hypothermia using transnasal high flow dry air. Ther Hypothermia Temp Manag. 2017;7(1):50–6.

Chapter 2
Hypoxia-Primed Stem Cell Transplantation in Stroke

Zheng Zachory Wei, James Ya Zhang, and Ling Wei

Abstract Spanning the past decade, stem cell research has made rapid progress, and stem cell transplantation in stroke has emerged as a promising treatment. Clinical applications of the cell-based therapy can benefit from the protective mechanisms of ischemic/hypoxic preconditioning. Genetic engineering techniques have been applied to the development of novel stem cell lines and augmenting the differentiation potency of different stem cells, which may ultimately provide far-reaching applications for translational studies, as well as developmental and pathological models.

Keywords Mesenchymal stromal cells · Neural progenitor cells · Pluripotent stem cells · HIF-1α · Hypoxic preconditioning

Spanning the past decade, stem cell research has made rapid progress, and stem cell transplantation in stroke has emerged as a promising treatment. Clinical applications of the cell-based therapy can benefit from the protective mechanisms of ischemic/hypoxic preconditioning. Genetic engineering techniques have been applied to the development of novel stem cell lines and augmenting the differentiation potency of different stem cells, which may ultimately provide far-reaching applications for translational studies, as well as developmental and pathological models.

Hypoxic and ischemic models are widely used in the research and development of new drugs and clinical therapy. Hypoxic and ischemic conditioning induced by a sublethal stimulus is an adaptive effect that confers enhanced

---

Z. Z. Wei · J. Y. Zhang
Department of Anesthesiology, Emory University School of Medicine, Atlanta, GA, USA

L. Wei (✉)
Department of Anesthesiology, Emory University School of Medicine, Atlanta, GA, USA

Department of Neurology, Emory University School of Medicine, Atlanta, GA, USA
e-mail: lwei7@emory.edu

© Springer Nature Switzerland AG 2019
J. Chen et al. (eds.), *Therapeutic Intranasal Delivery for Stroke and Neurological Disorders*, Springer Series in Translational Stroke Research,
https://doi.org/10.1007/978-3-030-16715-8_2

resistance to subsequent injuries that would have otherwise been lethal. Conditioning treatments on remote tissues or organs (remote conditioning) demonstrate great therapeutic efficacy with high translational potential. Transplantation of hypoxia-preconditioned cells is one of the feasible strategies that incorporate hypoxic tolerance in clinical applications. Hypoxic preconditioning and stem cell therapy display tremendous synergistic benefits in preclinical and clinical studies [1].

Ischemic/hypoxic preconditioning involves many endogenous defense mechanisms to induce cellular tolerance and therapeutic potentials. Preconditioning triggers include hypoxia/anoxia or exposure to agents such as apelin, carbon monoxide, cobalt protoporphyrin, diazoxide, erythropoietin (EPO), hydrogen dioxide (H_2O_2), heat shock protein (Hsp), hydrogen sulfide (H_2S), insulin-like growth factor-1 (IGF-1), isoflurane, lipopolysaccharide, and stromal-derived factor-1 (SDF-1).

After an effective preconditioning strategy, intranasal delivery of stem cells following ischemic stroke can be delivered into the brain within 24 h of stroke onset. Cells are able to reach the ischemic cortex and deposit outside of blood vessels as early as 1.5 h after administration. In hypoxia-primed stem cell transplantation for stroke, the transplanted preconditioned cells have a multitude of superior attributes, including: (a) enhanced survival and migration to replace damaged tissue, (b) suppression of inflammatory cytokines and downregulation of host immune responses against the allograft, (c) increased trophic factors and stimulation of regenerative healing to promote recovery. Notably, transplantation of preconditioned cells improved homing and integration to the lesion site. There is also greater integration of stem cells due to enhanced maturation and differentiation and the potential ability to promote host angiogenesis, arteriogenesis, neurogenesis, and synaptogenesis.

Stem cell and progenitor cell-based therapies using mesenchymal stem/stromal cells (MSC), endothelial progenitor cells (EPC), hematopoietic stem cells (HSC), oligodendrocyte progenitor cells (OPC), pluripotent stem cells (PSC) and c-kit+ cell population have been under extensive pre-clinical and clinical investigations for a variety of disorders.

Pluripotent stem cells PSCs, such as embryonic stem cells (ESC), are able to make cells from all three germ layers and have the potential to generate any cell/tissue for regeneration. Regardless of their cell source, PSC are capable of self-renewal and give rise to multiple specialized cell types in vitro and after transplantation. Human cell lines derived from ESC, as well as the creation of adult-induced pluripotent stem cells (iPSC) that allow for autologous applications for disease treatments, are most promising in stem cell therapy [2].

This chapter mainly discusses the potential applications of ESCs derived from the inner mass of blastocysts, iPSC reprogrammed from somatic cells, and adult MSCs available for intranasal treatment delivery. Generally, preconditioning of these cell types enhances cell adhesion and increased differentiation into vasculature. Hypoxia may promote neural differentiation of the above stem/progenitor cells. In addition, ESCs and iPSC are being combined with gene-editing techniques, which enable not only enhanced cell replacement, trophic support, drug delivery, immunomodulatory and anti-inflammatory effects, but also checkpoints and quality control functions for many on-going clinical trials. Genome editing greatly expands the understanding of pathological processes by studying cellular/disease models, as well as human cells and tissue, in which the programmable nucleases can directly correct or introduce genetic mutations. Compared to traditional drug therapy, therapeutic genome editing strategies provide an alternative method to treat both genetic diseases and acquired diseases that have genetic associations.

Strokes are devastating disorders that have complex pathophysiology that arise from a primary hypoxic insult. The initial insult causes a dysfunction of energy metabolism followed by massive cell death, glutamate excitotoxicity, free radical damage, reactive gliosis, activation of apoptotic cascades, acute and chronic inflammation, and other pathological pathways [3]. Hypoxia preconditioning applied to stem cells have been shown to enhance resistance to those injurious insults. The hypoxia-primed stem cells can increase pro-survival and anti-inflammatory signals, hypoxia-inducible factor (HIF)-1, trophic/growth factors, protein kinase B (Akt), extracellular signal-regulated kinase, glycogen synthase kinase-3β, matrix metalloproteinase-2, survivin, and B-cell lymphoma 2 (Bcl-2). HIF-1α and HIF-1β are nuclear factors with central roles in hypoxic/ischemic preconditioning and neuroprotection against ischemic injury. As a low-oxygen sensor, its translocation and activation in the nucleus results in production of several downstream genes such as CC chemokine receptor-7 (CCR-7), C-X-C chemokine receptor type 4 (CXCR-4), EPO, lactate dehydrogenase A, c-Met, matrix metalloproteinase-9 (MMP-9), pyruvate dehydrogenase kinase-1, sodium-calcium exchanger-1, uncoupling protein-2, and vascular endothelial growth factor (VEGF). HIF-1α stabilization also induces activation of protein kinase C (PKC) through mitochondrial mechanisms. PKC activates nuclear factor-kappa B (NF-κB) signaling, further enhances antioxidant and anti-apoptotic genes such as manganese superoxide dismutase (MnSOD) and Bcl-2, and promotes secretion of brain-derived neurotrophic factor (BDNF), FGF, and VEGF [4].

In adult animals after fetal tracheal occlusion, the blood pressure drops to an extremely low level within several minutes, triggering severe physiological responses. During these conditions, respiratory rate and cardiovascular activity enter an adaptive state. In many moderate to severe ischemic events, oxygen balance and collateral blood flow are controlled to be compensatory responses and the protective mechanisms are activated. Preconditioning harnesses these physiological responses in order to artificially provide equal and sometimes greater benefits than these organic controls. Sublethal ischemic events enhance tolerance to lethal ischemia in the brain and other organs. After preconditioning, neuroplasticity and hypoxic/ischemic adaptation can be induced, which involves changes in physiology, neurochemical, and neuroelectrophysiological properties [5].

2.1 Stem Cells for Intranasal Cell Therapy

Some recent improvements include: (a) *Transplantation of lineage-committed, fully-differentiated cells*. Generation of pure and differentiated specific type of cells are still needed. To eliminate the risk of tumorigenesis, quality control has been applied to the genetic properties of transplanted cells. Suicidal gene knock-in cells are ready for the elimination of potentially tumorigenic cells. (b) *Development for high-quality stable cell lines*. Standard operating procedures are developed to detect genetically-unstable cells and cancer stem cells. (c) *Continuous control of transplanted cells*. Technological advances and increasing corroboration from human studies allow the continuous monitoring of the exogenous cells and preventing tumor formation. (d) *Reducing transplant rejection*. The concern is that the cells after transplantation might trigger delayed transplant rejection. This concern is partially resolved now by utilizing autologous MSCs from the host's bone marrow, adipose tissue, and others, and differentiating iPSC reprogrammed from the host's own somatic cells. Alternatively, MSCs are low immunogenicity cells and show immunosuppressive effects after transplantation.

Ischemia/hypoxia and reactive oxygen species (ROS) are common players in tumorigenesis, stemness of cancer stem cells, and tumor progression [6]. The niche of cancer stem cells identified within many types of tumors or hematological cancers demonstrates hypoxic environment and low ROS conditions. Enforced tumoral expression of CD24, CD133, erythropoietin (EPO), HIF-1/MMPs, Janus kinase (JAK)/signal transducer and activator of transcription (STAT), Kruppel-like factor 4 (KLF-4), NANOG, nestin, Notch signaling, octamer-binding transcription factor 4 (OCT-4), SNAIL1, sodium calcium exchangers, transforming growth factor-1 (TGF-1), mothers against decapentaplegic homolog-4 (SMAD-4), and VEGF are found under the hypoxic lower-ROS conditions. HIF-1 and VEGF are two antitumoral angiogenic targets (e.g. anthracycline chemotherapy), which facilitate mobilization of circulating progenitors to the tumor angiogenesis. Hypoxia also induces stemness via reprogramming. The involved molecules in MSCs are fibroblast growth factor-2 (FGF-2), miRNA-302, NANOG, Notch-1, and OCT-4.

Stem cells provide developmental and pathological models. Based on cell potency and cell types, stem cells are classified as totipotent cells, naïve pluripotent stem cells, primed pluripotent cells, and tissue-specific multipotent stem cells. (a) *Totipotent cells*. Characteristic of the zygote, early blastomeres, and further reprogrammed/extended PSCs develop into all tissues, including extra-embryonic tissue. (b) *Naïve stem cells*. In a ground state, they harbor the prerequisite potential to differentiate into all embryonic lineages and develop into chimeric blastocysts. They possess high clonogenicity and do not carry specification markers. Naïve stem cells display greater levels of pluripotency marker proteins, including OCT-4, NANOG, SOX-2, KLF-2, and KLF-4. (c) *Primed pluripotent cells. e.g. human ESC*. They do not produce chimeras, express an FGF-5 specification marker and have low clonogenicity. Primed pluripotent cells lose KLF-2 and KLF-4 expression. (d) *Tissue-*

specific multipotent stem cells. They have the least differentiating potency among all stem cell types, with the ability to form tissue-specific cell types.

Cell sources in stem cell therapy may include: (a) *ESC*. They are a useful tool for exploring early embryonic development, modeling pathological processes of diseases, and developing therapeutics through drug discovery and potential regenerative medical treatments. ESC have very high differentiation efficiency into various transplantable progenitors/precursors and terminally differentiated neuronal and glial cell types, including cortical glutamatergic, striatal γ-aminobutyric acid gamma aminobutyric acid (GABA)-ergic, forebrain cholinergic, midbrain dopaminergic, serotonergic, and spinal motor neurons, as well as astrocytes and oligodendrocytes. (b) *iPSC*. They are pluripotent cells that are artificially de-differentiated from adult somatic cells by several transcription factors or small-molecule compounds, which are obtained from patients, and then banked and stored. They harbor much less ethical concern and opposition, and they minimize the risk of immune system rejection [7]. They are amenable as donor cells for cell replacement therapy, disease modeling, and drug screening. (c) *MSC*. They are from the bone marrow, adipose tissue and some other tissues. In stroke therapy, they show immunomodulatory and anti-inflammatory effects. Clinical trials demonstrate safety and feasibility of MSC transplantation in acute and chronic stroke with no tumorigenicity reported following cellular transplantation [8, 9]. (d) *Fetal NPC*. They differentiate into functional neurons and multiple types of neuroglia, allowing for the identification of specific neurodevelopmental processes related to the pathophysiology of developmental disorders. Fetal hNPC were investigated in clinical trials for the treatment of spinal cord injury (SCI) and age-related macular degeneration (AMD), but the safety has not yet been verified.

2.2 Conditioning Medicine and Cell Survival Mechanisms

Hypoxia, ischemia, or limited oxygen levels in different parts of the body all induce systemic changes under some microenvironmental conditions. For example, individuals in mountainous regions show adaptation to a lower oxygen level to maintain normal physiological functions for plateau residents at higher elevations. Neuronal activities consume a large amount of oxygen and glucose for maintenance of normal brain activities. Ischemic/hypoxic insults are more likely to have a greater deleterious effect within the brain. They are also more likely to cause a greater extent of damage in adult brains, as compared with embryos and newborns. In the body, the bone marrow cells survive well in the physiologically hypoxic conditions (1–6% O_2 in the bone marrow) and are potentially homing to the ischemic/hypoxic regions.

HIF-1 is a critical mediator in ischemia/hypoxia and ROS-induced responses. Under hypoxia, HIF-1 is involved in the activation of cytokines/chemokines, transcription factors and microRNAs (e.g. miRNAs-34a, 210, 214) for cell survival, metabolic adaptation, mitophagy, and mitochondrial biogenesis, and regulation of neurotrophic/angiogenic factors. HIF-1 isoforms upregulate β-catenin transcription and activate multiple proteins, including the activator protein 1, aryl hydrocarbon receptor (AhR), AhR nuclear translocator, bone morphogenetic protein (Bmp), cAMP response element-binding protein (CREB), cystathionine γ-lyase, cAMP-1-activated exchange protein Epac-1, forkhead box O3 (Foxo3), and elevation in hypoxia response element, MMPs, and sex-determining region Y box (SOX)-1. They also increase the expression of glucose-6-phosphate transporter, glucose transporter 1/3. HIF isoforms are coactivator of cell growth and autophagy regulator mechanistic target of rapamycin (mTOR), parkinsonism associated deglycase PARK7, multifunctional protein pyruvate kinase isozymes M1/M2 (PKM1/2).

2.3 Mitochondrial Mechanisms in Stem Cell and Stroke Treatment

Preconditioning results in better survival of bone marrow mesenchymal stromal cells and neural progenitor cells in vitro and/or after transplantation. This is particularly relevant for cell therapy because the survivability of transplanted cells is the primary issue after the cells are transplanted into the ischemic brain. Mitochondrial adaptation is one of the important protective mechanisms in preconditioned stem cells and stroke treatment [10, 11].

Mitochondrial ROS production triggered by H_2O_2, H_2S, and/or CO in ischemic brains and transplanted MSCs can induce ischemic/hypoxic tolerance mechanisms. Heme oxygenase-1 inducer cobalt protoporphyrin IX (CoPP) induces generation of endogenous CO and increases H_2O_2 to trigger the tolerance. Other antioxidant gene mediators include COX-2, Nrf2, and stanniocalcin-1. These protein/enzymes in the transplanted cells, as well as in the surrounding tissue, can retain survival signals, maintain cellular ion homeostasis, and regulate the balance between oxidative stress and glycolytic metabolism in mitochondria.

Decreased energy demands prevail under hypoxic and ischemic conditions as a compensatory response. However, severe ischemia-induced massive neuronal cell death, and endothelial/extracellular matrix (ECM) damage causes disruption of the blood-brain barrier. Ischemia/hypoxia-upregulated heat shock proteins (Hsp) including Hsp70 and Hsp90 can inhibit the mitochondrial release of second mitochondria-derived activator of caspase and prevent activation of caspase-9 and caspase-3. Hsp90 and Hsp70 may form a complex with Cx43 and facilitate the translocase of the outer membrane 20 (TOM20)-mediated translocation of Cx43 onto inner mitochondrial membranes. Hypoxic conditioning-induced HIF-1α stabilization reduces oxidative phosphorylation, leads to the opening of mitochondrial K_{ATP} channels and activates PKC.

2.4 Post-ischemic Flow Recovery

Reduction of blood oxygen and glucose levels, or an ischemic event caused by occlusion of blood vessels in the brain impact neuronal cell survival and neural function. Conditioning shows great benefits of neuroprotection and local cerebral blood flow recovery. Manipulation of hypoxic preconditioning, perconditioning, and postconditioning within the sublethal range in physiological and pathological conditions all show priming effects of improving the tolerance of cells, tissues, and the whole body from recent, on-going, and future insults.

Paracrine release of VEGF and EPO stimulate endogenous arteriogenesis, angiogenesis, and neurogenesis after ischemic stroke. HO-1/CO and SIRT1/eNOS/NO that regulate the cerebral vasodilation may also contribute to collateral circulation and flow recovery. Many genes of angiogenesis and arteriogenesis may be involved in the stem cell benefits and post-ischemic flow recovery (Table 2.1).

Hypoxic conditions induce the mobilization of endogenous stem cells. An ischemic insult to the cortex markedly increases SDF-1 in the ischemic region, a chemoattractant for directional migration of neuroblasts expressing CXCR-4. Hypoxia can also induce migration in various types of cells, including BMSC, cardiac SCA-1+ progenitors, ESCs/iPSC, NSC, and some tumor cells. The SDF-1/CXCR-4 axis and hypoxia are mediators for MSCs/EPC migration in the bone marrow, the peripheral blood, and many other organs. Activated SDF-1/CXCR-4 axis and monocyte chemoattractant protein 1 (MCP-1)/C-C chemokine receptor type 2 (CCR2) play important roles in migration of NPC and EPC after ischemic stroke, which direct the migrating neuroblasts to the infarct region for regeneration of the neurovascular network. Similarly, as treatment for myocardial infarction, hypoxia-induced upregulation of CXCR-4 in CD34+ stem/progenitor cells facilitated recruitment of donor CD34+ cells to the heart to protect against ischemia-reperfusion injury.

Mobilization of stem cells from the bone marrow demonstrates great therapeutic potentials. Bmp, EPO, granulocyte colony stimulating factor (G-CSF), and interleukin-10 (IL-10) mobilize endogenous bone marrow cells from the bone marrow, increase the homing and differentiation of NSC (originated from the neurogenic niches within the brain) and MSCs into the peri-infarct regions, and exert neuroprotective effects to promote stroke recovery and mitigate stroke damage. G-CSF mobilizes CD34+ hematopoietic stem cells and reduces microglial activation. Fasudil, an inhibitor of Rho kinase, is used to increase the G-CSF level for mobilization. In chronic hypoxia secondary to pulmonary hypertension, when migratory adaptation to SDF-1 and cell adhesion are significantly inhibited, hypoxic EPCs with both upregulated VEGFR-2+ and CXCR-4+ are insufficient for vascular remodeling. Enhancement of EPO/EPOR is demonstrated to attenuate hypoxia-induced pulmonary hypertension, while EPOR (−/−) mice display failed mobilization of EPCs to pulmonary endothelium for repair [12]. MMPs and natural MMP inhibitors are involved in ECM stabilization, glial activation, and regulation of migratory factors for stem cells and some other cells.

Table 2.1 Expression and mechanism of angiogenesis and arteriogenesis genes in stem cells

Gene family	Gene	UniProtKB ID	Cell type	Tissue source	Related cell	Functions
Peptidase M2 type	Ace	P12821	HSC	BM	–	Vascular morphogenesis
G-protein coupled receptor	Ackr3	P25106	NSC MSC	SVZ BM	Microglia	Cell migration, chemotaxis
	Adgrg1	Q9Y653	HSC	BM	–	VEGF-A production
	Ccr2	P41597	HSC NSC	BM SVZ	Monocyte	Cell migration, adhesion
	Cxcr3	P49682	NSC	SVZ, DG	–	Endothelial cell proliferation, survival, and chemotaxis
Hepatokine	Angptl3	Q9Y5C1	HSC	BM	–	Endothelial cell adhesion, and migration
Ankyrin SOCS box family	Asb5	Q8WWX0	–	–	Myogenic progenitor	Arteriogenesis
ATPase α/β chains family	Atp5f1b	P06576	–	–	–	Endothelial cell migration
–	Bcas3	Q9H6U6	–	–	–	Endothelial cell migration
TGF-β family	Bmp4	P12644	MSC NSC	BM SVZ, DG	Astroglia	Endothelial tube morphogenesis, endothelial cell proliferation, differentiation, migration
	Nodal	Q96S42	–	–	Macrophage	Angiogenesis
	Tgfb2	P61812	NPC	–	Neuron	Endothelial cell capillary morphogenesis, vascular smooth muscle cell proliferation, and migration
–	Bmper	Q8N8U9	HSC	BM	–	Endothelial cell activation, proliferation, migration
Glycosyltransferase family	B4galt1, C1galt1	P15291, Q9NS00	–	–	–	Cell adhesion, chemotaxis, cell survival

Collagen calcium-binding EGF domain-containing	Ccbe1	Q6UXH8	–	–	–	Endothelial cell migration
CD34 family	Cd34	P28906	HSC, MSC	BM	–	Endothelial cell proliferation, and adhesion
CMGC Ser/Thr PTK	Cdk5	Q00535	NSC	DG	Microglia	Endothelial cell migration, lamellipodia formation
	Mapk14	Q16539	NSC	SVZ, DG	Microglia	Angiogenesis
Carcinoembryonic antigen	Ceacam1	P13688	NPC	–	–	Endothelial cell differentiation, migration, sprouting angiogenesis, Arteriogenesis
β-Catenin family	Ctnnb1	P35222	NSC MSC, HSC	SVZ, DG BM	Microglia	Endothelial tube morphogenesis, cell differentiation
DimethylarginaseFamily	Ddah1	O94760	NPC	–	–	Regulation of arterial blood pressure
–	Dll1	O00548	NPC HSC	– BM	Myogenic progenitor	Sprouting angiogenesis, cell migration
RTK	Epha1, Epha2, Epha4, Ephb1, Ephb2, Ephb3, Ephb4	P21709, P29317, P54764, P54762, P29323, P54753, P54760	NSC	DG	Myogenic progenitor, neuron	Sprouting angiogenesis, endothelial cell proliferation, adhesion
Eph receptor interacting protein	Efna1, Efnb2, Efna3	P20827, P52799, P52797	–	–	–	
RTK	Flt1/ VEGFR1	P17948	HSC	BM	Macrophage	Sprouting angiogenesis, cell survival, migration, and chemotaxis

(continued)

Table 2.1 (continued)

Gene family	Gene	UniProtKB ID	Cell type	Tissue source	Related cell	Functions
Heparin binding growth factor	Fgf2, Fgf10	P09038, O15520	iPSC	–	VSMC	Endothelial cell proliferation, chemotaxis, sprouting angiogenesis
			HSC	BM		
Tyr PTK	Fgfr1, Fgfr2	P11362, P21802	NSC	OB, SVZ, DG	Neuron	Endothelial cell proliferation, migration, Chemotaxis
	Tie1, Tek/Tie2	P35590, Q02763	EPC	Blood		Sprouting angiogenesis; promote vascular stability
Forkhead box TF	Foxc1	Q12948	HSC, MSC	BM	–	Endothelial cell migration
	Foxo1, Foxo4, Foxp1	Q12778, P98177, Q9H334	NSC	Embryo		Endothelial cell proliferation, tube formation, migration
–	Gata2	P23769	HSC	Blood	Neuron	Endothelial cell proliferation
GRB2/Sem-5/DRK family	Grb2	P62993	NPC	–	–	Endothelial cell migration
Histone deacetylase family	Hdac5	Q9UQL6	NSC	DG	Neuron	Sprouting angiogenesis, cell migration
High mobility group proteins	Hmga2	P52926	HSC	BM	Myogenic progenitor	Cell proliferation
			NSC	SVZ		
Antennapedia homeobox	Hoxa7	P31268	MSC, HSC	BM	–	Cell differentiation
bHLH TF	Id3	Q02535	NSC	SVZ, DG	Astroglia	Gliovascular coupling
	Myc	P01106	iPSC	–	–	Sprouting angiogenesis
	Tal1	P17542	HSC	BM	–	Form microvasculature
–	Jag1	P78504	NSC	SVZ, DG	–	Sprouting angiogenesis
Jumonji-C domain containing protein	Jmjd6	Q6NYC1	MSC	Adipose	–	Sprouting angiogenesis

TCF/LEF family	Lef1	Q9UJU2	MSC	Adipose, BM	–	Sprouting angiogenesis
			NSC	DG		
Lysyl oxidase	Loxl2	Q9Y4K0	NPC	–	–	Endothelial cell proliferation, migration, sprouting angiogenesis
–	Naa15	Q9BXJ9	–	–	–	Regulation of vascular permeability
Basic leucine zipper TF	Atf2, Nfe2l2	P15336, Q16236	HSC	BM	Microglia, astroglia	Endothelial cell migration, survival
			NSC	SVZ		
Type-1 TM	Notch1	P46531	NSC	SVZ, DG	Microglia, astroglia, myogenic progenitor	Endothelial cell fate commitment, cell differentiation, gliovascular coupling
			MSC, HSC	BM		
CCN	Nov/Ccn3	P48745	HSC	BM	VSMC, neuron	Cell adhesion, chemotaxis, cell survival
Nuclear hormone receptor	Nr2e1	Q9Y466	NSC	SVZ, DG	Microglia, astroglia	Cell cycle progression
Plexin family	Plxdc1	Q8IUK5	–	–	–	Endothelial cell capillary morphogenesis
PI3K/PI4K	Pik3ca	P42336	–	–	Neuron	Endothelial cell migration, vasculogenesis
TKL Ser/Thr PTK	Raf1	P04049	HSC	BM	Macrophage, neuron	Cell motility, proliferation
Su(H) family	Rbpj	Q06330	–	–	Astroglia	Endothelial cell fate commitment and specification
–	Runx1, Runx2	Q01196, Q13950	NSC	SVZ, DG	Astroglia	Endothelial cell proliferation
			MSC	BM		

(continued)

Table 2.1 (continued)

Gene family	Gene	UniProtKB ID	Cell type	Tissue source	Related cell	Functions
–	Serpine1	P05121	MSC	BM	Astroglia	Endothelial cell survival
Histone lysine methyltransferase	Setd2	Q9BYW2	NSC	–	Myogenic progenitor	Vascular remodeling
			HSC	BM		
Secreted frizzled-related protein	Sfrp2	Q96HF1	HSC	BM	Neuron	Hematopoietic stem cell proliferation
			NSC	–		
–	Shb	Q15464	HSC	BM	Neuron	Endothelial cell proliferation, and differentiation
–	Shc1	P29353	NSC	SVZ	–	Endothelial cell migration, sprouting angiogenesis
Hedgehog	Shh	Q15465	NSC	SVZ, DG	Microglia, astroglia	Chemotaxis, vasculogenesis
Sirtuin protein deacetylases	Sirt1	Q96EB6	NSC	DG	Microglia, astroglia	Endothelial cell migration, proliferation
Dwarfin/SMAD family	Smad1	Q15797	NSC	SVZ	Neuron, myogenic progenitor	Endothelial cell proliferation, sprouting angiogenesis
			MSC, HSC	BM		
–	Srf	P11831	HSC	BM	Neuron, myogenic progenitor	Sprouting angiogenesis, cell migration
STAT TF	Stat3	P40763	NSC	SVZ, DG	Neuron, myogenic progenitor	Endothelial cell proliferation
			MSC	BM		
Sry-type HMG box	Sox17	Q9H6I2	NSC	–	–	Endothelial cell differentiation, vascular morphogenesis
	Sox18	P35713	MSC	BM		Endothelial cell migration, establish blood-brain barrier
RTase family	Tert	O14746	NSC	DG	–	Endothelial cell survival

Class I aminoacyl tRNA synthetase	Wars	P23381	–	–	Shear stress
Wingless/integrated	Wnt5a, Wnt7a, Wnt7b	P41221, O00755, P56706	HSC	BM	Endothelial cell proliferation, migration, neurovascular coupling
			NSC	SVZ, DG, OB	Neuron
CCCH-type ZFP	Zc3h12a	Q5D1E8	NPC	–	Macrophage
			MSC	UC, BM	Endothelial cell differentiation, gliogenesis

BM bone marrow, *DG* dentate gyrus, *EPC* endothelial progenitor cell, *HSC* hematopoietic stem cell, *iPSC* induced pluripotent stem cell, *MSC* mesenchymal stromal cell, *NPC* neural progenitor cell, *NSC* neural stem cell, *SVZ* subventricular zone, *TF* transcription factor, *TM* transmembrane, *OB* olfactory bulb, *PTK* protein kinase, *RTK* receptor tyrosine protein kinase, *UC* umbilical cord, *ZFP* zinc finger protein

Preconditioning with growth factors including FGF-2, glial cell line-derived neurotrophic factor (GDNF), IGF-1, SDF-1, and transforming growth factor-alpha (TGF-α), enhances paracrine effects. Upregulated factors may include angiopoietin-1, BDNF, EPO, FGF-2, GDNF, hepatocyte growth factor, MMP-2, placental growth factor, SDF-1, and VEGF. Treatment effects include: reduction of neurotoxicity and cell apoptosis, promotion of angiogenesis, neurogenesis, and synaptogenesis, and attenuation of functional and pathophysiological decline. Secreted factors also benefit the engraftment of stem/progenitor cells by enhancing cell survival, differentiation, and integration.

Enhancement of migration and homing of transplanted cells are relevant to the efficacy of cell-based therapy. Homing and promoted paracrine activity of endogenous stem cells as well as transplanted cells greatly contribute to flow recovery [13]. Plasma levels of SDF-1α and VEGF significantly increase during the subacute phases of ischemic stroke. Increased VEGF and SDF-1 in peripheral blood are involved in recruitment of EPC.

SDF-1/CXCR-4/CXCR-7 axis are well-known players in many cellular, physiological, and pathological processes, such as cell proliferation, migration, chemotaxis, inflammation, neurogenesis, angiogenesis, and hematopoiesis. In the peri-infarct regions of ischemic stroke, SDF-1 increases and interacts with CXCR-4/CXCR-7+ cells, leading to neurogenesis and neurovascular repairs. Upregulation of CXCR-4 and hepatocyte growth factor receptor, and activation of extracellular signal-regulated kinase (ERK), PI3 kinase-AKT, or PLCγ-PKC, MMP-2, and MMP-9 in MSCs by preconditioning insults enhances the migration and homing.

Circulating VEGF, TNF-α, and IL-8 also impact the recruitment of c-Kit+/Tie-2+ EPCs, CD34+ HSCs and MSCs towards the infarcted area. VEGF-A is an apoptosis inhibitor of vascular smooth muscle cells controlled by TGF-β/SMAD-3 signaling [14]. TGF-β signaling and TGF-β family genes such as Bmp4, Nodal, and Tgfb2 promote proliferation of endothelial cells and vascular smooth muscle cells for angiogenesis and arteriogenesis. VEGFR-1/Flt1 and VEGFR-2/Flk1 are essential for the mobilization and sprouting angiogenesis. Many chemokine and angiogenic genes are upregulated after hypoxic induction on bone marrow-derived hemangioblasts, which promote differentiation toward endothelial lineage and promote neovascularization. These include: Sonic Hedgehog (SHH), miRNA-31, miRNA-132 and miRNA-720 may be involved in EPC-mediated angiogenesis and neovascularization induced by angiopoietin-1, Eph family receptor-interacting protein B2 (ephrin B2), HIF-1α, methyl-CpG-binding protein 2, Ras-GTPase-activating protein, SDF-1, and VEGF. Angiotensin II pretreatment activated the AT1R/HIF-1α/ACE axis in rat BMSC and promoted VEGF production and the angiogenic response. Some negative regulators include: adhesion G protein-coupled receptors, angiopoietin-2/4, miRNA-377, Krueppel-like factor 4, leukemia inhibitory factor, Rho-associated protein kinases, semaphorins, and many others.

2.5 Stem Cells and Neuroplasticity

During the long-term period following ischemic stroke, there are neurodegenerative processes with gradual losses of neurons and neural connections. These include the interactions between neurons, astrocytes, microglial cells, oligodendrocytes, and stem/progenitor cells. To achieve the goal for cell-based repairs, cell type-specific commitment and differentiation protocols are developed [15].

Enhanced migration of endogenous NPC from the lateral ventricle subventricular zone (SVZ) towards the lesion sites increased the potential for cortical regeneration and repair. Wnt signaling players such as Wnt-3a, Wnt-5a, and Wnt-7a/b promote the neuroblast migration and differentiation contributing to neurogenesis, angiogenesis and neurovascular coupling. For example, intranasal delivery of Wnt-3a to the ischemic brain shows neuroprotection and greatly enhances neurogenesis, angiogenesis, and flow recovery [16, 17]. Another related factor in hypoxia preconditioned MSCs is Wnt-4, which modulates axonal growth. NSC and NPC in the adult brain proliferate and migrate from the two regenerative niches, the forebrain SVZ and the subgranular zone (SGZ) of the dentate gyrus in the hippocampus. Some other molecules that are shown to promote angiogenesis and arteriogenesis may also activate neural stem cells and neural progenitor cells stimulate regenerative processes such as endogenous neurogenesis. These genes include but are not limited to: Ackr3, Atf2, Bmp4, Ccr2, Ctnnb1, Mapk14, Nfe2l2, Fgfr1, Fgfr2, Hmga2, Id3, Jag1, Notch1, Nr2e1, Runx1, Runx2, Shc1, Shh, Smad1, and Stat3.

Regenerative neuronal circuits after stroke in rodent models include: (a) the interneurons originating from migrating neuroblasts along the SVZ and rostral migratory stream, toward the peri-infarct regions, striatum and the olfactory bulb; (b) the SGZ neuroblasts differentiate and integrate within the hippocampus. Lipid accumulation, perturbation of the microenvironmental fatty acid metabolism, and inhibition of Wnt signaling and VEGFR signaling, have been shown to suppress the homeostatic and regenerative functions of NSC and NPC. Inhibitor of CXCR-4 such as Plerixafor (also known as age-related macular degeneration, AMD-3100) significantly increases VEGFR-2-positive cells in the peripheral blood, elevates SDF-1 levels, and promotes blood vessel formation in an ischemic flap model. Co-culture of neurons with SDF-1-secreting olfactory ensheathing cells after oxygen-glucose deprivation (OGD) treatment, an in vitro method of hypoxic/ischemic preconditioning, showed enhanced neurite outgrowth. In addition, SDF-1-overexpressed NPC derived from iPSC show enhanced axonal and synaptic growth, and increased numbers of NeuN/BrdU and Glut-1/BrdU co-labeled cells in peri-infarct regions after transplantation.

Hypoxia affects the NSC phenotype, cell differentiation, and regenerative repair activity. The proliferation and self-renewal of NSC are maintained under hypoxic conditions. HIF-1α overexpression promotes NSC proliferation and differentiation after intracerebral hemorrhage and hypoxic/ischemic injury. Other related factors include: angiotensin II (Ang II), angiotensin-converting enzyme (ACE), Ang II receptor Type 1 (AT1R), EPO/EPOR, Leukemia inhibitory factor (LIF)/pSTAT-3,

LIF receptor (LIFR)/glycoprotein (gp130), Neurogenin-1/BMP-4, Notch-1, atypical protein kinase C/CREB-binding protein pathway, SOX-2, and VEGF/VEGFR. HIF-1α is also required for neural stem cell maintenance in the adult mouse SVZ as well as for mouse embryonic stem cells toward a neural lineage.

Glial progenitor cells in the SVZ and the white matter proliferate and differentiate after ischemia, which is further augmented by exogenous treatments, including epidermal growth factor (EGF), EPO, GDNF, memantine, and uridine diphosphoglucose glucose, and by the enhancement of BDNF, SHH, VEGF, and Wnt signaling in the SVZ or in transplanted cells. MSCs may produce some other factors including Ang-2, FGF, GDF-5, HGF, IGF-1, LIF, MCP-1, SCF, TGF-β, TIMP-1, TIMP-2, and TSP-1 [18].

Enhanced neuronal maturation, oligogenesis, and synaptogenesis by preconditioning is an effective approach that is logically related to the regenerative efficacy of stem cell-based therapies. Hypoxic preconditioning increases secretion of growth factors and upregulation of their cognate receptors, such as CXCR-4, as well as greater expression of neuregulin-1 isotype β-1/ErbB4, neurofilament, stem cell antigen 1, and synaptophysin. Neuregulin-1 isotype β-1/ErbB4 signaling protects OPCs during and after a hypoxic event in the white matter.

2.6 Inflammation, Immune Responses, and Regeneration

The increased inducible nitric oxide synthase (iNOS), an antioxidant genes, are involved in the regulation of cell fate in the inflammatory microenvironment. Preconditioned cells show inhibitory effects on cyclooxygenase production, and they reduce inflammation by releasing anti-inflammatory factors. In MSCs, downregulated expression of pro-inflammatory cytokines/chemokines and receptors include CC3, CC5, CC17, CCL4, CXCR3, CXCL10, IL-1β, IL-6, TNF-α, IFN-γ, and OX-42. Hypoxia-preconditioned MSCs suppress microglial activation and gliosis in the ischemic brain. MSCs inhibit T-cells and natural killer (NK) cells, and they reduce immune responses by decreasing proliferation of immunocytes. Activation of the transplanted cells may further suppress inflammatory and immune responses in host tissues.

Intranasal delivery of preconditioned stem cells improves motor recovery and promotes regenerative activities, which contribute to greater improvement in coordination skills, neuropsychiatric, and cognitive functions. In the investigations on ischemic stroke models, there are significantly more NeuN-positive and NeuN/BrdU-colabeled neurons, MBP myelination, and Glut1-positive and Glut1/BrdU-colabeled cells in the ischemic core and peri-infarct regions. Co-transplanting stem cells may promote revascularization via paracrine effects. Advanced methods are bringing new approaches for enhanced cell quality/adaptability and improved transplantation therapy for human diseases.

Stem cell transplantation after ischemic stroke stimulates angiogenesis, ameliorates ischemia-hypoxia, and provides nutrient support. Intranasal delivery of BMSC in the acute phase exerts neuroprotective benefits after ischemia. Hypoxic-preconditioned BMSC and neural progenitors showed significant increases in the survival of transplanted cells, homing to the lesion sites, neuronal differentiation, and functional benefits after stroke. Hypoxia-treated hMSC contain a secretive enrichment of trophic factors that provide a suitable preconditioning strategy for enhanced differentiation of endogenous NPC after transplantation.

Disease modeling and drug screening studies using iPSC allow for greater experimental interrogation and convenience compared to transgenic animal models [19, 20]. iPSC provide for the potential use of reprogrammed somatic cells from the patient to establish disease-relevant phenotypes in vitro and to simulate and recapitulate the molecular signatures of pathogenesis during an early stage [21]. A combination therapy with iPSC and genome editing is proposed as a new therapeutic paradigm to introduce protective mutations, to correct deleterious mutations, to eliminate the antigenic/immunogenic signals in the iPSC, or to destroy foreign viral DNAs in the human body. Genome editing of iPSC uses programmable nucleases including (a) *transcription activator-like effector nucleases* (TALENs). In the absence of exogenous template DNA, the programmable nucleases create a double strand break (DSB) in desired regions, but due to the error-prone no homologous end joining (NHEJ) mechanism of re-ligation, an insertion/deletion (indel) mutation is frequently created at the DSB site. (b) *Clustered regularly interspaced short palindromic repeats* (CRISPR)/*Cas9* technology. To ablate the triple repeats, a pair of single-guide RNAs was applied to target both sides during the expansion. Other than NHEJ, the high-fidelity homology-directed repair (HDR)-based mechanism of genome editing is studied to treat deleterious loss-of-function mutations. HDR-based genome editing provides an exogenous repair template of a single-stranded oligodeoxynucleotide and a donor plasmid to correct a mutated allele to be wild type. It could also integrate therapeutic transgenes into a genomic safe harbor site.

Modified iPSC and its derived multiple cell types will be another useful cell source for treating stroke and many other neurological disorders. Here are unmet translational gaps that warrant further investigation: (a) *Identification of MSC derived from iPSC*. (b) *Intranasal delivery of hypoxia-preconditioned MSC derived from iPSC*. (c) *Angiogenesis and arteriogenesis after transplantation*. (d) *Blood flow recovery*. (e) *Neuroplasticity after transplantation*.

References

1. Wei L, et al. Stem cell transplantation therapy for multifaceted therapeutic benefits after stroke. Prog Neurobiol. 2017;157:49–78.
2. Fox IJ, et al. Stem cell therapy. Use of differentiated pluripotent stem cells as replacement therapy for treating disease. Science. 2014;345(6199):1247391.
3. Eltzschig HK, Eckle T. Ischemia and reperfusion--from mechanism to translation. Nat Med. 2011;17(11):1391–401.
4. Fraisl P, Aragones J, Carmeliet P. Inhibition of oxygen sensors as a therapeutic strategy for ischaemic and inflammatory disease. Nat Rev Drug Discov. 2009;8(2):139–52.

5. Li S, et al. Preconditioning in neuroprotection: from hypoxia to ischemia. Prog Neurobiol. 2017;157:79–91.
6. Liu Y, et al. The microenvironment in classical Hodgkin lymphoma: an actively shaped and essential tumor component. Semin Cancer Biol. 2014;24:15–22.
7. Yamanaka S. Induced pluripotent stem cells: past, present, and future. Cell Stem Cell. 2012;10(6):678–84.
8. Eckert MA, et al. Evidence for high translational potential of mesenchymal stromal cell therapy to improve recovery from ischemic stroke. J Cereb Blood Flow Metab. 2013;33(9):1322–34.
9. van Velthoven CT, et al. Mesenchymal stem cell transplantation attenuates brain injury after neonatal stroke. Stroke. 2013;44(5):1426–32.
10. Yu SP, Wei Z, Wei L. Preconditioning strategy in stem cell transplantation therapy. Transl Stroke Res. 2013;4(1):76–88.
11. Dirnagl U, Becker K, Meisel A. Preconditioning and tolerance against cerebral ischaemia: from experimental strategies to clinical use. Lancet Neurol. 2009;8(4):398–412.
12. Satoh K, et al. Important role of endogenous erythropoietin system in recruitment of endothelial progenitor cells in hypoxia-induced pulmonary hypertension in mice. Circulation. 2006;113(11):1442–50.
13. Wei ZZ, et al. Enhanced neurogenesis and collaterogenesis by sodium danshensu treatment after focal cerebral ischemia in mice. Cell Transplant. 2018;27(4):622–36.
14. Shi X, et al. TGF-beta/Smad3 inhibit vascular smooth muscle cell apoptosis through an autocrine signaling mechanism involving VEGF-A. Cell Death Dis. 2014;5:e1317.
15. Tabar V, Studer L. Pluripotent stem cells in regenerative medicine: challenges and recent progress. Nat Rev Genet. 2014;15(2):82–92.
16. Wei ZZ, et al. Neuroprotective and regenerative roles of intranasal Wnt-3a administration after focal ischemic stroke in mice. J Cereb Blood Flow Metab. 2018;38(3):404–21.
17. Matei N, et al. Intranasal wnt3a attenuates neuronal apoptosis through Frz1/PIWIL1a/FOXM1 pathway in MCAO rats. J Neurosci. 2018;38(30):6787–801.
18. Ranganath SH, et al. Harnessing the mesenchymal stem cell secretome for the treatment of cardiovascular disease. Cell Stem Cell. 2012;10(3):244–58.
19. Dolmetsch R, Geschwind DH. The human brain in a dish: the promise of iPSC-derived neurons. Cell. 2011;145(6):831–4.
20. Sandoe J, Eggan K. Opportunities and challenges of pluripotent stem cell neurodegenerative disease models. Nat Neurosci. 2013;16(7):780–9.
21. Merkle FT, Eggan K. Modeling human disease with pluripotent stem cells: from genome association to function. Cell Stem Cell. 2013;12(6):656–68.

Chapter 3
Therapeutic Potential of Intranasal Drug Delivery in Preclinical Studies of Ischemic Stroke and Intracerebral Hemorrhage

Qian Li, Claire F. Levine, and Jian Wang

Abstract Stroke is the leading cause of death in the United States. Both ischemic stroke and hemorrhagic stroke cause high mortality and morbidity, but few effective treatments are available clinically. The lack of therapeutics stems partially from the severe nature of stroke, which causes a cascade of events from low blood flow to neuronal death, to inflammation that can last for weeks. However, the biggest obstacle to treatment is the blood-brain barrier (BBB), which blocks passage of most systemically delivered drugs into the brain. Noninvasive intranasal (IN) delivery enables drugs to rapidly bypass the BBB and minimizes exposure to the peripheral system. Thus, IN is a promising delivery route for treating central nervous system disorders, including stroke. In this chapter, we introduce the mechanism of neurologic damage after stroke and describe the advantages and disadvantages of IN delivery. Additionally, we systematically review preclinical animal studies that have used IN to treat ischemic and hemorrhagic stroke and briefly discuss the status of IN delivery in clinical trials for stroke.

Keywords Intranasal delivery · Ischemic stroke · Intracerebral hemorrhagic stroke

3.1 Introduction

Stroke is the second leading cause of death worldwide and is thought to have affected approximately 6.7 million people in 2012 (http://www.who.int/mediacentre/factsheets/fs310/en/). The most common type is ischemic stroke, which accounts for 85% of all stroke cases [1]. It occurs when a blood vessel in the brain is blocked by a clot or by severe narrowing. The result is critically reduced blood flow. Hemorrhagic stroke, which accounts for 15–20% of stroke cases, is caused by arteries leaking blood or rupturing in the brain [2]. The resulting hematoma puts pressure

Q. Li · C. F. Levine · J. Wang (✉)
Department of Anesthesiology and Critical Care Medicine, The Johns Hopkins University School of Medicine, Baltimore, MD, USA
e-mail: qli34@jhmi.edu; clevine4@jhmi.edu; jwang79@jhmi.edu

© Springer Nature Switzerland AG 2019
J. Chen et al. (eds.), *Therapeutic Intranasal Delivery for Stroke and Neurological Disorders*, Springer Series in Translational Stroke Research, https://doi.org/10.1007/978-3-030-16715-8_3

on brain cells and causes cell death. Hypertension, trauma, blood-thinning medications, and aneurysms increase the chances of hemorrhage. The third type of stroke is the transient ischemic attack (TIA), which is caused by a clot or debris that temporarily interrupts arterial blood flow [3]. According to the Centers for Disease Control and Prevention (CDC), over a third of people who experience a TIA will experience a major stroke within a year if not treated (https://www.cdc.gov/stroke/types_of_stroke.htm).

The gold standard for treating ischemic stroke is tissue plasminogen activator (tPA) [4], which dissolves clots and improves blood flow to the brain. Another treatment option is mechanical thrombectomy, in which large clots are removed by a stent retriever [5]. For hemorrhagic stroke, the only option is to stop the bleeding surgically [5]. Not only are treatment options limited, the treatment window for stroke is also very short. For instance, tPA works for patients only within 3 h of onset [5], and inclusion criteria for tPA treatment are very strict (e.g., 18 years or older; clinical diagnosis of ischemic stroke with a measurable neurologic deficit; and <3 h since time of onset) [6]. Thus, more effective treatment methods and drugs are urgently needed.

Although stroke has been researched in preclinical studies for decades, and a variety of chemical, peptide, and genetic therapies have shown promise in animal models, few therapies have had success clinically. The obstacle to translating preclinical success into clinical success is not just a matter of species differences. It is also a physiologic obstacle known as the blood-brain barrier (BBB), which prevents passage of toxic materials and infectious microbes, but also blocks entry of potentially therapeutic drugs. Translational science needs a way to bypass the BBB and effectively deliver drugs to the brain [7]. Intranasal (IN) delivery is a promising new delivery route that was developed by Frey in 1989 [98]. With this approach, drug is delivered directly from the nasal cavity into the central nervous system (CNS) and bypasses the BBB [8]. IN delivery reduces systemic exposure, which thereby reduces side effects and boosts delivery efficiency. Even more important, it is rapid and noninvasive compared with other routes of delivery [8]. In this chapter, we introduce and review IN delivery and its application to translational and clinical stroke studies.

3.2 Mechanisms of Neurologic Damage after Stroke

When blood flow in the brain is critically reduced, ischemic neurons start to die from oxygen and glucose deprivation [7]. ATP shortage disturbs the function of energy-dependent membrane ion pumps (sodium, calcium, and potassium), leading to an influx of calcium ions [7]. The overload of intracellular calcium triggers the release of excitatory neurotransmitter glutamate, and leads to neuronal swelling. Excess calcium also generates free radicals and reactive oxygen species (ROS) and activates calcium-dependent enzymes (calpain, endonucleases, ATPases, etc.) causing apoptotic and/or necrotic cell death [7]. The acute damage to neuronal cells lasts several hours. In the sub-acute and chronic phases, immune cells (microglia and other infiltrating cells) are recruited to the infarcted region and initiate inflammatory reactions. Pro-inflammatory cytokines are upregulated and induce cell death for the next several days to weeks, gradually increasing the infarct volume [7]. Therapeutic strategies can include removing glutamate, reducing ROS, decreasing inflammation, and modifying the polarization of microglia/macrophages [7, 9, 10].

In the case of hemorrhagic stroke, cells begin to die immediately after bleeding begins because of hematoma expansion and the consequent increase in intracranial pressure [11]. Secondary damage after intracerebral hemorrhage (ICH) results mainly from the leaked blood components and immune cell reactions [12–14]. Plasma-derived thrombin induces expression of prostaglandin E2 EP3 receptor and contributes to a pro-inflammatory response via RhoA [12], while blood degradation product hemoglobin releases free iron and causes neuronal cell death by increasing ROS. Activated resident microglia and astrocytes, as well as infiltrating leukocytes and macrophages, release cytokines, chemokines, prostaglandins, proteases, ferrous iron, and other immunoactive molecules that cause inflammation and secondary neuronal death [11, 15]. Neurons undergo cell death from mechanisms such as apoptosis, necrosis, and autophagy in the acute and sub-acute phases after ICH (until 7 days), but the inflammatory response lasts for over 2 weeks and subsequently increases brain edema, BBB damage, and lesion volume [11]. Therapeutic strategies include decreasing neuronal ROS and iron toxicity, reducing inflammation, and modifying the polarization of microglia/macrophages [12–14, 16, 17].

3.3 Intranasal Drug Delivery

Three structural components separate brain and blood flow: the BBB, the blood-cerebrospinal-fluid barrier (BCSFB), and the arachnoid membrane [7]. Considering that the surface area of the BBB is more than 1000-fold greater than that of the BCSFB or arachnoid membrane, the BBB is the primary barrier to delivery of drugs to the CNS [7]. The BBB separates circulating blood from the brain's extracellular fluid with a layered structure—endothelial cells, capillary basement membrane, pericytes, and astrocyte endfeet tightly surround the capillaries and form a sheath [18]. The permeability of this barrier is highly selective. Small lipophilic compounds, such as O_2 and CO_2, can diffuse freely across the BBB along their concentration gradients via tight junctions between the cells [18, 19]; larger molecules require active transport and carrier-mediated transport, such as efflux pumps, endocytosis, paracellular transport, transcellular passive diffusion, and receptor-mediated transport [18, 20].

Owing to the highly selective nature of the BBB, classic drug delivery methods such as intravenous (IV), intraperitoneal (IP), and intra-arterial (IA) injections are quite inefficient. Most of the administered drug is excreted or metabolized in the kidneys, lungs, and liver. Indeed, <1% of drug administered through an IV injection reaches the CNS [21]. Moreover, these systemic routes of administration expose the circulation and healthy tissue to the drug, thereby increasing the risk of adverse effects [8, 18]. Another option is to bypass the BBB entirely by using intracranial or intracerebroventricular injection. However these routes are invasive and costly, require hospitalization, are inherently risky, and cannot be used for long-term treatment [18]. IN, in contrast, provides a safe, rapid, and convenient delivery method that is also low-cost and noninvasive. By bypassing the systemic circulation, IN delivery also reduces the risk of side effects.

IN-administered therapeutics may reach the CNS via direct and indirect pathways. With indirect transport, molecules first enter the lymphatic and vascular circulation, after which they must cross the BBB to reach the brain [22]. Given the rapid speed with which drugs enter the brain after IN delivery (within minutes), increasing evidence indicates that drugs may actually be absorbed via the direct pathway, specifically through the olfactory and trigeminal pathways [8, 22]. Those nerves connect nasal mucosa and the brain directly. Substrates may gain entry into the CNS by leaking between the nasal epithelium cells through the intercellular spaces; alternatively, they may need to be transported across the nasal epithelium by channels or pores. Lipid-soluble and smaller molecules may enter via energy-independent membrane diffusion (Fig. 3.1) [8, 22].

Drugs, peptides, proteins, solid nanoparticles, and even cells can be delivered intranasally [22, 23]. Certain reagents known as adsorption enhancing compounds can be used to increase the intranasal bioavailability of protein and peptide drugs. Cyclodextrins and albumin are most commonly used for this purpose [22]. Several

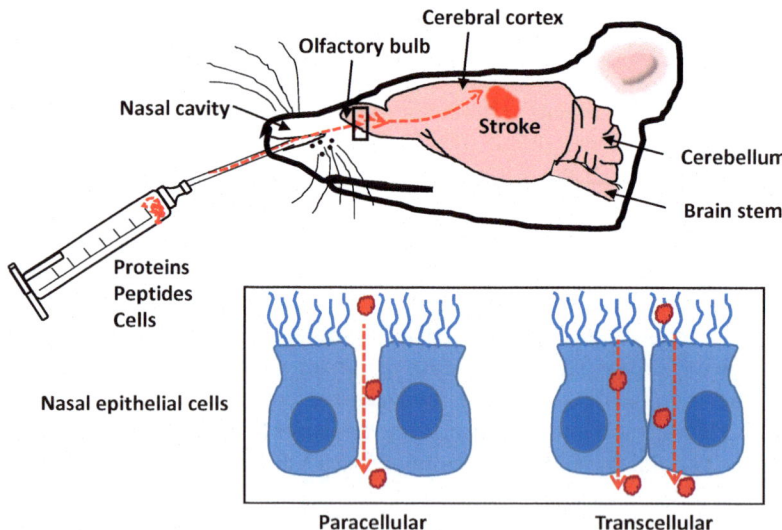

Fig. 3.1 Anatomical overview of nasal cavity and olfactory tissue in a rodent, illustrating the intranasal route. Proteins, peptides, and cells delivered intranasally enter the brain and perihematoma area via the olfactory bulb. Delivered substrates pass through nasal epithelial cells by paracellular or transcellular pathways, depending on their size and physiochemical properties

forms of cyclodextrin can be used, including α (six glucose residues), β (seven glucose residues), and γ (eight glucose residues). Interestingly, each of these cyclodextrins has a different contribution pattern: The α-cyclodextrin increases uptake by the olfactory bulb and decreases uptake by the occipital cortex, striatum, and whole brain; the β-cyclodextrin increases uptake by all brain regions except the striatum and olfactory bulb; and hydro-β-cyclodextrin increases uptake by the thalamus and decreases uptake by the striatum [24]. Albumin was also able to help target peptides to various regions of the brain in a study of IN leptin delivery. Co-administration of albumin with leptin decreased serum levels of leptin, possibly by reducing its clearance from the brain. Albumin also increased leptin uptake into the hypothalamus, a region that would enhance its effect on feeding and cognition, but decreased leptin uptake into non-target regions, such as the cerebellum [25]. Using these reagents to target different brain regions would greatly benefit the efficiency and decrease potential side effects.

3.4 IN Treatment for Ischemic Stroke

Various proteins and genes have been investigated as therapeutic agents in experimental animal models of ischemic stroke. We review several of the most studied therapies here.

3.4.1 *Insulin-Like Growth Factor-1 (IGF-1)*

IGF-1, a hormone similar in molecular structure to insulin, plays an important role in growth during development and continues to have anabolic effects during adulthood [26]. Some investigators have reported that serum IGF-1 level can be used as a potential predictor of ischemic stroke outcome clinically [27–32]. Moreover, several studies have shown that IGF-1 protects against stroke in experimental animal models when delivered by invasive methods [33–35]. Additionally, in a clinical study, Donath et al. [36] found that administering IGF-1 intravenously to patients with chronic heart failure is safe and improves cardiac performance by afterload reduction and possibly by positive inotropic effects. Taken together, these data suggest that IGF-1 therapy may hold promise for ischemic stroke in future clinical trials.

Translational researchers have been attempting to deliver IGF-1 intranasally since 2001 [37–40]. Liu et al. [39] showed that 150 μg of IGF-1 delivered IN reduced infarct volume and improved motor, sensory, reflex, and vestibulomotor functions in rats after 2 h of middle cerebral artery occlusion (MCAO). In a later study, they showed that even 75 μg of IGF-1 (225 μg total IGF-1 over 48 h) was sufficient to reduce infarct volume, hemispheric swelling, and neurologic deficits

[40]. They also showed that the neuroprotection of IN-delivered IGF-1 was effective up to 6 h post-MCAO, and that pretreating (1 h before MCAO) or post-treating (2 h after MCAO) was beneficial when infusing IGF-1 intracerebroventricularly [38, 41, 42].

Another promising drug, erythropoietin, has been shown to have protective effects when delivered IN together with IGF-1 [37]. Fletcher et al. [37] showed that IN administration of erythropoietin plus IGF-1 reduced stroke volume within 24 h and improved neurologic function in mice up to 90 days after MCAO. More importantly, they labeled IGF-1 and erythropoietin with ^{125}I and found that concentrations of each peaked at 60–120 min after IN delivery versus 240–360 min for the other delivery systems (subcutaneous, IV, and IP) [37]. Moreover, brain concentrations of erythropoietin were significantly higher after IN delivery than after other delivery methods, and the ischemic brain had higher concentrations than did normal brain [37].

3.4.2 Erythropoietin

Systemic erythropoietin is a hematopoietic growth factor that regulates red blood cell production and inhibits erythrocytic progenitor apoptosis [43]. Independent of its hematopoietic actions, erythropoietin and its variants are directly neuroprotective in various ischemic models [44]. IV administration of human recombinant erythropoietin (rHu-EPO) was shown to significantly reduce infarct volume and provide notable neurologic and clinical improvement after ischemia in patients; however, systemic exposure caused side effects by elevating hematocrit [45]. Bypassing the periphery with IN delivery could reduce or potentially eliminate such side effects.

Merelli et al. [46, 47] showed that IN delivery of rHu-EPO had a positive effect in a $CoCl_2$-induced focal brain hypoxia rat model. They found that IN rHu-EPO improved spontaneous motor activity of hypoxic rats significantly in the open field and rotarod tests. Additionally, Gao et al. [48] showed that in Mongolian gerbils with brain injury induced by unilateral permanent ischemia, IN-delivered recombinant human neuronal erythropoietin (Neuro-EPO) decreased mortality by reducing delayed neuronal death.

To enhance the translational possibility of rHu-EPO and Neuro-EPO, investigators have tested different dosages and frequencies in normal brain (mouse, Mongolian gerbil, and nonhuman primate) and in ischemic brain (mouse, rat, and Mongolian gerbil) for safety and efficacy [44]. Total administration of 4800 IU erythropoietin was safe for rodents, and 5750 IU/kg was safe for nonhuman primates; even one dose of 12 IU was effective in a focal ischemic stroke model, and 4 days' treatment with 249.4 IU was effective in a global ischemic stroke model [44].

3.4.3 Osteopontin

Osteopontin is a large secreted glycoprotein with an arginine, glycine, and aspartate (RGD) motif that is expressed in brain neurons and microglia [49]. By binding to and signaling through cellular integrin receptors expressed on most CNS cells, it regulates adhesion, migration, differentiation, survival, and repair [49]. Studies have shown that osteopontin provides neuroprotection in animal models of ischemic stroke through integrin receptor signaling [50]. Doyle et al. [49] investigated the effects of IN osteopontin as well as thrombin-cleaved osteopontin in a mouse MCAO model. Thrombin cleavage helps to expose the RGD sequence and likely increases the specificity of integrin binding. The investigators found that thrombin-cleaved osteopontin provided significantly more effective neuroprotection than did native osteopontin. Moreover, they found that IN administration of osteopontin, thrombin-treated osteopontin, or osteopontin peptides successfully targeted ischemic brain and decreased infarct volume compared to vehicle treatment. Jin et al. [51] showed that RGD-containing osteopontin icosamer peptide (OPNpt20) offered a robust neuroprotective effect by suppressing the upregulation of inducible nitric oxide synthase (iNOS) and other inflammatory markers. Endogenous $\alpha v\beta 3$ integrin appeared to play a critical role in suppressing iNOS. Notably, peptide containing a mutated RGD motif failed to decrease infarct volume [51]. The same group also showed that using gelatin nanoparticles to carry osteopontin extended the treatment window to 6 h post-MCAO from 3 h when osteopontin was used alone [52].

3.4.4 Transforming Growth Factor (TGF)

The TGF family consists of two classes: TGF-α and TGF-β. TGF-α is the ligand of epidermal growth factor receptor, which activates a signaling pathway for cell proliferation, differentiation, and development. TGF-α is believed to be a prognostic biomarker in various tumors [53, 54]. It also has been shown to reduce infarct volume [55, 56], improve memory function [57], and enhance neurogenesis and angiogenesis [58] in animal models of ischemic stroke. However, TGF-α can be delivered only by invasive means because it does not cross the BBB. Therefore, Guerra-Crespo et al. [59] bound the TGF-α molecule to polyethylene glycol (PEG) to increase its stability and delivered it intranasally in a rat MCAO model. They found that the IN-delivered PEGylated TGF-α reduced behavioral deficits by inducing proliferation of neural progenitors and their migration to the damaged striatum.

The cytokine TGF-β1 modulates multiple functions, such as cell growth and differentiation, inflammation, and cell repair [60–62], and was shown to protect neurons from hypoxic/ischemic injuries [63, 64]. Ma et al. [65] showed that among mice subjected to transient MCAO, those treated with IN TGF-β1 had significantly better neurologic function and smaller infarct volume than did those in a saline-treated control group. TGF-β1 treatment suppressed apoptosis in the ipsilateral striatum and increased BrdU⁺ cells in the subventricular zone and ipsilateral striatum. Importantly, IN delivery of TGF-β1 increased the neuronal commitment of neural progenitor cells [65].

3.4.5 Mesenchymal Stem Cells (MSCs)

Stem cell therapy holds promise for treating conditions such as cancer, spinal cord injury, and stroke [66]. Other than embryonic stem cells, bone marrow-derived MSCs (BMSCs) are the most studied stem cell type. Populations can be expanded extensively through cloning, and they have the capacity to differentiate into osteocytes, adipocytes, and chondrocytes under specific *in vitro* stimuli. These characteristics, plus their natural ability to migrate toward wound tissue and tumors, make them a perfect candidate for delivery of therapeutic reagents [66, 67].

IN delivery of one million BMSCs to neonatal rats or adult mice that had undergone stroke significantly improved neurologic, sensory, and social functions; reduced infarct volume; rescued neuronal death; and promoted angiogenesis, neurogenesis, and neurovascular repair [23, 68]. Additionally, preconditioning of BMSCs with hypoxia (HP-BMSCs) enhanced tolerance and regenerative properties [68]. Wei et al. [23] showed that preconditioned cells reached ischemic cortex and accumulated outside of the vasculature as early as 1.5 h after administration. Furthermore, administered HP-BMSCs boosted expression of proteins associated with migration, including CXC chemokine receptor type 4, matrix metalloproteinase (MMP)-2, and MMP-9 [23]. In another set of experiments, van Velthoven et al. [69] showed that MSCs that overexpress brain-derived neurotrophic factor (BDNF) reduced infarct volume, gray matter loss, and white matter loss with no significant difference from unaltered MSCs; however, MSC-BDNF further reduced rat pup motor deficits 14 days after MCAO compared to those in the MSC-treated group.

3.4.6 Other Factors and Reagents

Other well-studied reagents for treating ischemic stroke also have been shown to be safe and effective when delivered intranasally. For instance, 600 μg of IN tPA pushed the treatment window for ischemic stroke in rats to 7 days post-MCAO [70], and IN deferoxamine decreased infarct size in rats after MCAO without significantly changing blood pressure or heart rate [71]. Also of note, brain concentrations of deferoxamine in rats were significantly higher after 6-mg IN delivery (0.9–18.5 μM) than after 6-mg IV delivery (0.1–0.5 μM). Nerve growth factor was reported to accumulate in the hippocampus after IN delivery, improve vestibular function, and decrease infarct volume in rats with MCAO [72]. N6-cyclopentyladenosine, basic fibroblast growth factor, granulocyte colony-stimulating factor, exendin-4, interleukin-1 receptor antagonist, ginsenoside Rb1, progesterone, apelin, Xingnaojing microemulsion, caspase-9 inhibitor, and high-mobility group box 1 binding heptamer peptide have also shown beneficial effects toward ischemic injury [73–83].

3.5 IN Treatment for Intracerebral Hemorrhagic Stroke

Although ICH has been less well studied than ischemic stroke, some investigators have explored using IN delivery of therapeutics in ICH models. Sun et al. [84] investigated the use of IN-delivered rat HP-BMSCs in a mouse model of collagenase-induced ICH. They administered the HP-BMSCs at 3 and 7 days after ICH and found that the cells arrived at the ipsilateral cortex, perivascular spaces, and perihematoma region at 6 h after transplantation. Protein levels of glial cell-derived neurotrophic factor, vascular endothelial growth factor, and BDNF increased, as did the number of NeuN$^+$/BrdU$^+$ co-labeled cells in the perihematoma region, at 14 days post-ICH. Neurologic function of injured mice was significantly improved at days 14 and 21 post-ICH [84].

Quinpirole is another reagent that has been tested in an ICH model [85]. Quinpirole is a dopamine D2 receptor agonist that has been shown to play an important role in controlling innate immunity and inflammatory response after stroke [86, 87]. Zhang et al. [85] found that IN quinpirole (both 3 and 15 mg/kg), delivered at 1 h after ICH induced by autologous blood injection, improved modified Garcia score, improved forelimb placing score, and reduced brain water content at 1 day.

In a study conducted by Tsuchiyama et al. [88], nicotinamide adenine dinucleotide (NAD$^+$; 10 or 20 mg/kg) administered intranasally to mice with collagenase-induced ICH failed to show neuroprotection at 24 h, although IN NAD$^+$ did reduce infarct volume and improve neurologic function in a rat MCAO model [89].

3.6 Clinical Trials of IN Drug Delivery for Stroke

Belokoskova and colleagues conducted a series of clinical trials on IN delivery of V2 vasopressin receptor agonist 1-desamino-8-D-arginine vasopressin (DDAVP) in stroke patients. In two studies of 26 and 64 patients with different forms of stroke-induced aphasia, they found that IN delivery of DDAVP improved speech function in 79% of cases [90, 91]. The effects persisted during a 2-year follow-up period, and the neuropeptide optimized the activity of both the right and left cerebral hemispheres [92]. In another study of 40 stroke patients who were administered IN DDAVP for 1.5–2 months, patients experienced decreases in both dysthymia and major depressive disorder for 0.5–1 year after the first course of therapy [93].

Currently in the US, three open clinical trials are in preparation or recruiting patients to study IN delivery of therapies for stroke. These include (1) a pilot study to investigate the role of IN insulin on post-stroke cognitive impairment (ClinicalTrials.gov identifier: NCT02810392); (2) a phase I/phase II study of IN bioactive factors (BFs) produced by autologous M2 macrophages in patients with organic brain syndrome, including brain ischemia (ClinicalTrials.gov identifier: NCT02957123); and (3) a phase I study to determine if IN E-selectin can prevent a second stroke in patients who have already experienced a stroke or TIA (ClinicalTrials.gov identifier: NCT00069069).

3.7 Future Directions and Conclusion

The use of intranasal delivery systems for proteins, peptides, and stem cells holds promise for treating CNS disorders. IN delivery has several advantages. First, it allows substrates to access the CNS directly without negotiating the BBB [94]. Consequently, the concentration of the substrate is much greater than it would be if administered through a systemic route. Second, it greatly minimizes exposure of peripheral organs to the substrates, thereby significantly decreasing potential circulatory toxicity and side effects [22]. Third, substrates arrive at the CNS much more rapidly when they are administered IN than when they are administered peripherally, usually within minutes. This advantage allows the use of some drugs that are degraded rapidly and might not otherwise be an option [95]. Finally, IN delivery is noninvasive, does not require hospitalization, and is very suitable for long-term treatment [22].

Of course, IN delivery has some disadvantages as well. For example it can induce a nasal reaction and may cause irritation of the cells and nerves in the nasal cavity [96]. Second, absorption across the nasal epithelium may be limited in patients with degradation of the nasal mucosa or if the surface area for absorption is small [22, 94, 96]. The distance a substrate must traverse to reach the target brain region can also be a limitation [94, 97]. In the future, modified molecules or molecular analogs should be studied as a means to increase delivery efficiency and specific brain region targeting. Finding ways to decrease irritation of the nasal cavity and increase drug penetration of the nasal mucosa represent additional study directions. Furthermore, animal studies are needed to explore the mechanism by which proteins and peptides are transported from the nasal cavity to the brain. An understanding of this process will enable scientists to search for ways to improve transport and increase targeting efficiency [22]. As the study of IN delivery increases in preclinical translational stroke research, it has the potential to revolutionize the approach for treatment of patients with stroke and other brain disorders.

References

1. Bushnell C, LD MC, Awad IA, Chireau MV, Fedder WN, Furie KL, Howard VJ, Lichtman JH, Lisabeth LD, Pina IL, Reeves MJ, Rexrode KM, Saposnik G, Singh V, Towfighi A, Vaccarino V, Walters MR, American Heart Association Stroke Council, Council on Cardiovascular and Stroke Nursing, Council on Clinical Cardiology, Council on Epidemiology and Prevention, Council for High Blood Pressure Research. Guidelines for the prevention of stroke in women: a statement for healthcare professionals from the American Heart Association/American Stroke Association. Stroke. 2014;45(5):1545–88. https://doi.org/10.1161/01.str.0000442009.06663.48.
2. Hemphill JC 3rd, Greenberg SM, Anderson CS, Becker K, Bendok BR, Cushman M, Fung GL, Goldstein JN, Macdonald RL, Mitchell PH, Scott PA, Selim MH, Woo D, American Heart Association Stroke Council, Council on Cardiovascular and Stroke Nursing, Council on Clinical Cardiology. Guidelines for the management of spontaneous intracerebral hemorrhage: a guideline for healthcare professionals from the American Heart Association/American Stroke Association. Stroke. 2015;46(7):2032–60. https://doi.org/10.1161/STR.0000000000000069.
3. Donnan GA, Fisher M, Macleod M, Davis SM. Stroke. Lancet. 2008;371(9624):1612–23. https://doi.org/10.1016/S0140-6736(08)60694-7.
4. Wardlaw JM, Murray V, Berge E, del Zoppo GJ. Thrombolysis for acute ischemic stroke. Cochrane Database Syst Rev. 2014;7:CD000213. https://doi.org/10.1002/14651858.CD000213.pub3.
5. Emberson J, Lees KR, Lyden P, Blackwell L, Albers G, Bluhmki E, Brott T, Cohen G, Davis S, Donnan G, Grotta J, Howard G, Kaste M, Koga M, von Kummer R, Lansberg M, Lindley RI, Murray G, Olivot JM, Parsons M, Tilley B, Toni D, Toyoda K, Wahlgren N, Wardlaw J, Whiteley W, del Zoppo GJ, Baigent C, Sandercock P, Hacke W, Stroke Thrombolysis Trialists' Collaborative Group. Effect of treatment delay, age, and stroke severity on the effects of intravenous thrombolysis with alteplase for acute ischemic stroke: a meta-analysis of individual patient data from randomised trials. Lancet. 2014;384(9958):1929–35. https://doi.org/10.1016/S0140-6736(14)60584-5.
6. Miller DJ, Simpson JR, Silver B. Safety of thrombolysis in acute ischemic stroke: a review of complications, risk factors, and newer technologies. Neurohospitalist. 2011;1(3):138–47. https://doi.org/10.1177/1941875211408731.
7. Rhim T, Lee DY, Lee M. Drug delivery systems for the treatment of ischemic stroke. Pharm Res. 2013;30(10):2429–44. https://doi.org/10.1007/s11095-012-0959-2.
8. Hanson LR, Frey WH 2nd. Intranasal delivery bypasses the blood-brain barrier to target therapeutic agents to the central nervous system and treat neurodegenerative disease. BMC Neurosci. 2008;9(Suppl 3):S5. https://doi.org/10.1186/1471-2202-9-S3-S5.
9. Li Q, Han X, Wang J. Organotypic hippocampal slices as models for stroke and traumatic brain injury. Mol Neurobiol. 2016;53(6):4226–37. https://doi.org/10.1007/s12035-015-9362-4.
10. Zhang Z, Zhang Z, Lu H, Yang Q, Wu H, Wang J. Microglial polarization and inflammatory mediators after intracerebral hemorrhage. Mol Neurobiol. 2017;54(3):1874–86. https://doi.org/10.1007/s12035-016-9785-6.
11. Wang J. Preclinical and clinical research on inflammation after intracerebral hemorrhage. Prog Neurobiol. 2010;92(4):463–77. https://doi.org/10.1016/j.pneurobio.2010.08.001.
12. Han X, Lan X, Li Q, Gao Y, Zhu W, Cheng T, Maruyama T, Wang J. Inhibition of prostaglandin E2 receptor EP3 mitigates thrombin-induced brain injury. J Cereb Blood Flow Metab. 2016;36(6):1059–74. https://doi.org/10.1177/0271678X15606462.
13. Lan X, Han X, Liu X, Wang J. Inflammatory responses after intracerebral hemorrhage: from cellular function to therapeutic targets. J Cereb Blood Flow Metab. 2019;39:184–86. https://doi.org/10.1177/0271678X18805675.
14. Wang J, Dore S. Inflammation after intracerebral hemorrhage. J Cereb Blood Flow Metab. 2007;27(5):894–908. https://doi.org/10.1038/sj.jcbfm.9600403.

15. Wu H, Zhang Z, Hu X, Zhao R, Song Y, Ban X, Qi J, Wang J. Dynamic changes of inflammatory markers in brain after hemorrhagic stroke in humans: a postmortem study. Brain Res. 2010;1342:111–7. https://doi.org/10.1016/j.brainres.2010.04.033.
16. Wu H, Wu T, Li M, Wang J. Efficacy of the lipid-soluble iron chelator 2,2′-dipyridyl against hemorrhagic brain injury. Neurobiol Dis. 2012;45(1):388–94. https://doi.org/10.1016/j.nbd.2011.08.028.
17. Zhao X, Wu T, Chang CF, Wu H, Han X, Li Q, Gao Y, Li Q, Hou Z, Maruyama T, Zhang J, Wang J. Toxic role of prostaglandin E2 receptor EP1 after intracerebral hemorrhage in mice. Brain Behav Immun. 2015;46:293–310. https://doi.org/10.1016/j.bbi.2015.02.011.
18. Stockwell J, Abdi N, Lu X, Maheshwari O, Taghibiglou C. Novel central nervous system drug delivery systems. Chem Biol Drug Des. 2014;83(5):507–20. https://doi.org/10.1111/cbdd.12268.
19. Grieb P, Forster RE, Strome D, Goodwin CW, Pape PC. O2 exchange between blood and brain tissues studied with 18O2 indicator-dilution technique. J Appl Physiol (1985). 1985;58(6):1929–41.
20. Abbott NJ, Patabendige AA, Dolman DE, Yusof SR, Begley DJ. Structure and function of the blood-brain barrier. Neurobiol Dis. 2010;37(1):13–25. https://doi.org/10.1016/j.nbd.2009.07.030.
21. Pardridge WM. Drug transport across the blood-brain barrier. J Cereb Blood Flow Metab. 2012;32(11):1959–72. https://doi.org/10.1038/jcbfm.2012.126.
22. Meredith ME, Salameh TS, Banks WA. Intranasal delivery of proteins and peptides in the treatment of neurodegenerative diseases. AAPS J. 2015;17(4):780–7. https://doi.org/10.1208/s12248-015-9719-7.
23. Wei N, Yu SP, Gu X, Taylor TM, Song D, Liu XF, Wei L. Delayed intranasal delivery of hypoxic-preconditioned bone marrow mesenchymal stem cells enhanced cell homing and therapeutic benefits after ischemic stroke in mice. Cell Transplant. 2013;22(6):977–91. https://doi.org/10.3727/096368912X657251.
24. Nonaka N, Farr SA, Nakamachi T, Morley JE, Nakamura M, Shioda S, Banks WA. Intranasal administration of PACAP: uptake by brain and regional brain targeting with cyclodextrins. Peptides. 2012;36(2):168–75. https://doi.org/10.1016/j.peptides.2012.05.021.
25. Falcone JA, Salameh TS, Yi X, Cordy BJ, Mortell WG, Kabanov AV, Banks WA. Intranasal administration as a route for drug delivery to the brain: evidence for a unique pathway for albumin. J Pharmacol Exp Ther. 2014;351(1):54–60. https://doi.org/10.1124/jpet.114.216705.
26. Laron Z. Insulin-like growth factor 1 (IGF-1): a growth hormone. Mol Pathol. 2001;54(5):311–6.
27. De Geyter D, De Smedt A, Stoop W, De Keyser J, Kooijman R. Central IGF-I receptors in the brain are instrumental to neuroprotection by systemically injected IGF-I in a rat model for ischemic stroke. CNS Neurosci Ther. 2016;22(7):611–6. https://doi.org/10.1111/cns.12550.
28. De Geyter D, Stoop W, Sarre S, De Keyser J, Kooijman R. Neuroprotective efficacy of subcutaneous insulin-like growth factor-I administration in normotensive and hypertensive rats with an ischemic stroke. Neuroscience. 2013;250:253–62. https://doi.org/10.1016/j.neuroscience.2013.07.016.
29. Guan J, Williams C, Gunning M, Mallard C, Gluckman P. The effects of IGF-1 treatment after hypoxic-ischemic brain injury in adult rats. J Cereb Blood Flow Metab. 1993;13(4):609–16. https://doi.org/10.1038/jcbfm.1993.79.
30. Rizk NN, Myatt-Jones J, Rafols J, Dunbar JC. Insulin like growth factor-1 (IGF-1) decreases ischemia-reperfusion induced apoptosis and necrosis in diabetic rats. Endocrine. 2007;31(1):66–71.
31. Schabitz WR, Hoffmann TT, Heiland S, Kollmar R, Bardutzky J, Sommer C, Schwab S. Delayed neuroprotective effect of insulin-like growth factor-i after experimental transient focal cerebral ischemia monitored with mri. Stroke. 2001;32(5):1226–33.
32. Smith PF. Neuroprotection against hypoxia-ischemia by insulin-like growth factor-I (IGF-I). IDrugs. 2003;6(12):1173–7.
33. Denti L, Annoni V, Cattadori E, Salvagnini MA, Visioli S, Merli MF, Corradi F, Ceresini G, Valenti G, Hoffman AR, Ceda GP. Insulin-like growth factor 1 as a predictor of ischemic

stroke outcome in the elderly. Am J Med. 2004;117(5):312–7. https://doi.org/10.1016/j.amjmed.2004.02.049.
34. Johnsen SP, Hundborg HH, Sorensen HT, Orskov H, Tjonneland A, Overvad K, Jorgensen JO. Insulin-like growth factor (IGF) I, -II, and IGF binding protein-3 and risk of ischemic stroke. J Clin Endocrinol Metab. 2005;90(11):5937–41. https://doi.org/10.1210/jc.2004-2088.
35. Tang JH, Ma LL, Yu TX, Zheng J, Zhang HJ, Liang H, Shao P. Insulin-like growth factor-1 as a prognostic marker in patients with acute ischemic stroke. PLoS One. 2014;9(6):e99186. https://doi.org/10.1371/journal.pone.0099186.
36. Donath MY, Sutsch G, Yan XW, Piva B, Brunner HP, Glatz Y, Zapf J, Follath F, Froesch ER, Kiowski W. Acute cardiovascular effects of insulin-like growth factor I in patients with chronic heart failure. J Clin Endocrinol Metab. 1998;83(9):3177–83. https://doi.org/10.1210/jcem.83.9.5122.
37. Fletcher L, Kohli S, Sprague SM, Scranton RA, Lipton SA, Parra A, Jimenez DF, Digicaylioglu M. Intranasal delivery of erythropoietin plus insulin-like growth factor-I for acute neuroprotection in stroke. Laboratory investigation. J Neurosurg. 2009;111(1):164–70. https://doi.org/10.3171/2009.2.JNS081199.
38. Liu XF, Fawcett JR, Hanson LR, Frey WH 2nd. The window of opportunity for treatment of focal cerebral ischemic damage with noninvasive intranasal insulin-like growth factor-I in rats. J Stroke Cerebrovasc Dis. 2004;13(1):16–23. https://doi.org/10.1016/j.jstrokecerebrovasdis.2004.01.005.
39. Liu XF, Fawcett JR, Thorne RG, DeFor TA, Frey WH 2nd. Intranasal administration of insulin-like growth factor-I bypasses the blood-brain barrier and protects against focal cerebral ischemic damage. J Neurol Sci. 2001;187(1–2):91–7.
40. Liu XF, Fawcett JR, Thorne RG, Frey WH 2nd. Non-invasive intranasal insulin-like growth factor-I reduces infarct volume and improves neurologic function in rats following middle cerebral artery occlusion. Neurosci Lett. 2001;308(2):91–4.
41. Gluckman P, Klempt N, Guan J, Mallard C, Sirimanne E, Dragunow M, Klempt M, Singh K, Williams C, Nikolics K. A role for IGF-1 in the rescue of CNS neurons following hypoxic-ischemic injury. Biochem Biophys Res Commun. 1992;182(2):593–9.
42. Guan J, Waldvogel HJ, Faull RL, Gluckman PD, Williams CE. The effects of the N-terminal tripeptide of insulin-like growth factor-1, glycine-proline-glutamate in different regions following hypoxic-ischemic brain injury in adult rats. Neuroscience. 1999;89(3):649–59.
43. Fisher JW. Erythropoietin: physiology and pharmacology update. Exp Biol Med (Maywood). 2003;228(1):1–14.
44. Garcia-Rodriguez JC, Sosa-Teste I. The nasal route as a potential pathway for delivery of erythropoietin in the treatment of acute ischemic stroke in humans. ScientificWorldJournal. 2009;9:970–81. https://doi.org/10.1100/tsw.2009.103.
45. Ehrenreich H, Hasselblatt M, Dembowski C, Cepek L, Lewczuk P, Stiefel M, Rustenbeck HH, Breiter N, Jacob S, Knerlich F, Bohn M, Poser W, Ruther E, Kochen M, Gefeller O, Gleiter C, Wessel TC, De Ryck M, Itri L, Prange H, Cerami A, Brines M, Siren AL. Erythropoietin therapy for acute stroke is both safe and beneficial. Mol Med. 2002;8(8):495–505.
46. Merelli A, Caltana L, Girimonti P, Ramos AJ, Lazarowski A, Brusco A. Recovery of motor spontaneous activity after intranasal delivery of human recombinant erythropoietin in a focal brain hypoxia model induced by CoCl2 in rats. Neurotox Res. 2011;20(2):182–92. https://doi.org/10.1007/s12640-010-9233-8.
47. Merelli A, Caltana L, Lazarowski A, Brusco A. Experimental evidence of the potential use of erythropoietin by intranasal administration as a neuroprotective agent in cerebral hypoxia. Drug Metabol Drug Interact. 2011;26(2):65–9. https://doi.org/10.1515/DMDI.2011.007.
48. Gao Y, Mengana Y, Cruz YR, Munoz A, Teste IS, Garcia JD, Wu Y, Rodriguez JC, Zhang C. Different expression patterns of Ngb and EPOR in the cerebral cortex and hippocampus revealed distinctive therapeutic effects of intranasal delivery of Neuro-EPO for ischemic insults to the gerbil brain. J Histochem Cytochem. 2011;59(2):214–27. https://doi.org/10.1369/0022155410390323.

49. Doyle KP, Yang T, Lessov NS, Ciesielski TM, Stevens SL, Simon RP, King JS, Stenzel-Poore MP. Nasal administration of osteopontin peptide mimetics confers neuroprotection in stroke. J Cereb Blood Flow Metab. 2008;28(6):1235–48. https://doi.org/10.1038/jcbfm.2008.17.
50. Meller R, Stevens SL, Minami M, Cameron JA, King S, Rosenzweig H, Doyle K, Lessov NS, Simon RP, Stenzel-Poore MP. Neuroprotection by osteopontin in stroke. J Cereb Blood Flow Metab. 2005;25(2):217–25. https://doi.org/10.1038/sj.jcbfm.9600022.
51. Jin YC, Lee H, Kim SW, Kim ID, Lee HK, Lee Y, Han PL, Lee JK. Intranasal delivery of RGD motif-containing osteopontin Icosamer confers neuroprotection in the postischemic brain via alphavbeta3 integrin binding. Mol Neurobiol. 2016;53(8):5652–63. https://doi.org/10.1007/s12035-015-9480-z.
52. Joachim E, Kim ID, Jin Y, Kim KK, Lee JK, Choi H. Gelatin nanoparticles enhance the neuroprotective effects of intranasally administered osteopontin in rat ischemic stroke model. Drug Deliv Transl Res. 2014;4(5–6):395–9. https://doi.org/10.1007/s13346-014-0208-9.
53. Fanelli MF, Chinen LT, Begnami MD, Costa WL Jr, Fregnami JH, Soares FA, Montagnini AL. The influence of transforming growth factor-alpha, cyclooxygenase-2, matrix metalloproteinase (MMP)-7, MMP-9 and CXCR4 proteins involved in epithelial-mesenchymal transition on overall survival of patients with gastric cancer. Histopathology. 2012;61(2):153–61. https://doi.org/10.1111/j.1365-2559.2011.04139.x.
54. Tarhini AA, Lin Y, Yeku O, LaFramboise WA, Ashraf M, Sander C, Lee S, Kirkwood JM. A four-marker signature of TNF-RII, TGF-alpha, TIMP-1 and CRP is prognostic of worse survival in high-risk surgically resected melanoma. J Transl Med. 2014;12:19. https://doi.org/10.1186/1479-5876-12-19.
55. Justicia C, Perez-Asensio FJ, Burguete MC, Salom JB, Planas AM. Administration of transforming growth factor-alpha reduces infarct volume after transient focal cerebral ischemia in the rat. J Cereb Blood Flow Metab. 2001;21(9):1097–104. https://doi.org/10.1097/00004647-200109000-00007.
56. Justicia C, Planas AM. Transforming growth factor-alpha acting at the epidermal growth factor receptor reduces infarct volume after permanent middle cerebral artery occlusion in rats. J Cereb Blood Flow Metab. 1999;19(2):128–32. https://doi.org/10.1097/00004647-199902000-00002.
57. Alipanahzadeh H, Soleimani M, Soleimani Asl S, Pourheydar B, Nikkhah A, Mehdizadeh M. Transforming growth factor-alpha improves memory impairment and neurogenesis following ischemia reperfusion. Cell J. 2014;16(3):315–24.
58. Leker RR, Toth ZE, Shahar T, Cassiani-Ingoni R, Szalayova I, Key S, Bratincsak A, Mezey E. Transforming growth factor alpha induces angiogenesis and neurogenesis following stroke. Neuroscience. 2009;163(1):233–43. https://doi.org/10.1016/j.neuroscience.2009.05.050.
59. Guerra-Crespo M, Sistos A, Gleason D, Fallon JH. Intranasal administration of PEGylated transforming growth factor-alpha improves behavioral deficits in a chronic stroke model. J Stroke Cerebrovasc Dis. 2010;19(1):3–9. https://doi.org/10.1016/j.jstrokecerebrovasdis.2009.09.005.
60. Ishikawa M, Jin Y, Guo H, Link H, Xiao BG. Nasal administration of transforming growth factor-beta1 induces dendritic cells and inhibits protracted-relapsing experimental allergic encephalomyelitis. Mult Scler. 1999;5(3):184–91.
61. Mishra L, Derynck R, Mishra B. Transforming growth factor-beta signaling in stem cells and cancer. Science. 2005;310(5745):68–71. https://doi.org/10.1126/science.1118389.
62. Sporn MB, Roberts AB, Wakefield LM, Assoian RK. Transforming growth factor-beta: biological function and chemical structure. Science. 1986;233(4763):532–4.
63. Henrich-Noack P, Prehn JH, Krieglstein J. TGF-beta 1 protects hippocampal neurons against degeneration caused by transient global ischemia. Dose-response relationship and potential neuroprotective mechanisms. Stroke. 1996;27(9):1609–14; discussion 1615.
64. McNeill H, Williams C, Guan J, Dragunow M, Lawlor P, Sirimanne E, Nikolics K, Gluckman P. Neuronal rescue with transforming growth factor-beta 1 after hypoxic-ischaemic brain injury. Neuroreport. 1994;5(8):901–4.
65. Ma M, Ma Y, Yi X, Guo R, Zhu W, Fan X, Xu G, Frey WH 2nd, Liu X. Intranasal delivery of transforming growth factor-beta1 in mice after stroke reduces infarct volume and

increases neurogenesis in the subventricular zone. BMC Neurosci. 2008;9:117. https://doi.org/10.1186/1471-2202-9-117.
66. Li Q, Wijesekera O, Salas SJ, Wang JY, Zhu M, Aprhys C, Chaichana KL, Chesler DA, Zhang H, Smith CL, Guerrero-Cazares H, Levchenko A, Quinones-Hinojosa A. Mesenchymal stem cells from human fat engineered to secrete BMP4 are nononcogenic, suppress brain cancer, and prolong survival. Clin Cancer Res. 2014;20(9):2375–87. https://doi.org/10.1158/1078-0432.CCR-13-1415.
67. Pendleton C, Li Q, Chesler DA, Yuan K, Guerrero-Cazares H, Quinones-Hinojosa A. Mesenchymal stem cells derived from adipose tissue vs bone marrow: in vitro comparison of their tropism towards gliomas. PLoS One. 2013;8(3):e58198. https://doi.org/10.1371/journal.pone.0058198.
68. Wei ZZ, Gu X, Ferdinand A, Lee JH, Ji X, Ji XM, Yu SP, Wei L. Intranasal delivery of bone marrow mesenchymal stem cells improved neurovascular regeneration and rescued neuropsychiatric deficits after neonatal stroke in rats. Cell Transplant. 2015;24(3):391–402. https://doi.org/10.3727/096368915X686887.
69. van Velthoven CT, Sheldon RA, Kavelaars A, Derugin N, Vexler ZS, Willemen HL, Maas M, Heijnen CJ, Ferriero DM. Mesenchymal stem cell transplantation attenuates brain injury after neonatal stroke. Stroke. 2013;44(5):1426–32. https://doi.org/10.1161/STROKEAHA.111.000326.
70. Liu Z, Li Y, Zhang L, Xin H, Cui Y, Hanson LR, Frey WH 2nd, Chopp M. Subacute intranasal administration of tissue plasminogen activator increases functional recovery and axonal remodeling after stroke in rats. Neurobiol Dis. 2012;45(2):804–9. https://doi.org/10.1016/j.nbd.2011.11.004.
71. Hanson LR, Roeytenberg A, Martinez PM, Coppes VG, Sweet DC, Rao RJ, Marti DL, Hoekman JD, Matthews RB, Frey WH 2nd, Panter SS. Intranasal deferoxamine provides increased brain exposure and significant protection in rat ischemic stroke. J Pharmacol Exp Ther. 2009;330(3):679–86. https://doi.org/10.1124/jpet.108.149807.
72. Zhao HM, Liu XF, Mao XW, Chen CF. Intranasal delivery of nerve growth factor to protect the central nervous system against acute cerebral infarction. Chin Med Sci J. 2004;19(4):257–61.
73. Akpan N, Serrano-Saiz E, Zacharia BE, Otten ML, Ducruet AF, Snipas SJ, Liu W, Velloza J, Cohen G, Sosunov SA, Frey WH 2nd, Salvesen GS, Connolly ES Jr, Troy CM. Intranasal delivery of caspase-9 inhibitor reduces caspase-6-dependent axon/neuron loss and improves neurological function after stroke. J Neurosci. 2011;31(24):8894–904. https://doi.org/10.1523/JNEUROSCI.0698-11.2011.
74. Chen D, Lee J, Gu X, Wei L, Yu SP. Intranasal delivery of Apelin-13 is neuroprotective and promotes angiogenesis after ischemic stroke in mice. ASN Neuro. 2015;7(5). https://doi.org/10.1177/1759091415605114.
75. Dalpiaz A, Gavini E, Colombo G, Russo P, Bortolotti F, Ferraro L, Tanganelli S, Scatturin A, Menegatti E, Giunchedi P. Brain uptake of an anti-ischemic agent by nasal administration of microparticles. J Pharm Sci. 2008;97(11):4889–903. https://doi.org/10.1002/jps.21335.
76. Frechou M, Zhang S, Liere P, Delespierre B, Soyed N, Pianos A, Schumacher M, Mattern C, Guennoun R. Intranasal delivery of progesterone after transient ischemic stroke decreases mortality and provides neuroprotection. Neuropharmacology. 2015;97:394–403. https://doi.org/10.1016/j.neuropharm.2015.06.002.
77. Kim ID, Shin JH, Lee HK, Jin YC, Lee JK. Intranasal delivery of HMGB1-binding heptamer peptide confers a robust neuroprotection in the postischemic brain. Neurosci Lett. 2012;525(2):179–83. https://doi.org/10.1016/j.neulet.2012.07.040.
78. Lee JH, Kam EH, Kim JM, Kim SY, Kim EJ, Cheon SY, Koo BN. Intranasal administration of interleukin-1 receptor antagonist in a transient focal cerebral ischemia rat model. Biomol Ther (Seoul). 2017;25(2):149–57. https://doi.org/10.4062/biomolther.2016.050.
79. Lu T, Jiang Y, Zhou Z, Yue X, Wei N, Chen Z, Ma M, Xu G, Liu X. Intranasal ginsenoside Rb1 targets the brain and ameliorates cerebral ischemia/reperfusion injury in rats. Biol Pharm Bull. 2011;34(8):1319–24.

80. Sun BL, He MQ, Han XY, Sun JY, Yang MF, Yuan H, Fan CD, Zhang S, Mao LL, Li DW, Zhang ZY, Zheng CB, Yang XY, Li YV, Stetler RA, Chen J, Zhang F. Intranasal delivery of granulocyte colony-stimulating factor enhances its neuroprotective effects against ischemic brain injury in rats. Mol Neurobiol. 2016;53(1):320–30. https://doi.org/10.1007/s12035-014-8984-2.
81. Wen R, Zhang Q, Xu P, Bai J, Li P, Du S, Lu Y. Xingnaojing mPEG2000-PLA modified microemulsion for transnasal delivery: pharmacokinetic and brain-targeting evaluation. Drug Dev Ind Pharm. 2016;42(6):926–35. https://doi.org/10.3109/03639045.2015.1091471.
82. Zhang H, Meng J, Zhou S, Liu Y, Qu D, Wang L, Li X, Wang N, Luo X, Ma X. Intranasal delivery of exendin-4 confers neuroprotective effect against cerebral ischemia in mice. AAPS J. 2016;18(2):385–94. https://doi.org/10.1208/s12248-015-9854-1.
83. Zhao YZ, Lin M, Lin Q, Yang W, Yu XC, Tian FR, Mao KL, Yang JJ, Lu CT, Wong HL. Intranasal delivery of bFGF with nanoliposomes enhances in vivo neuroprotection and neural injury recovery in a rodent stroke model. J Control Release. 2016;224:165–75. https://doi.org/10.1016/j.jconrel.2016.01.017.
84. Sun J, Wei ZZ, Gu X, Zhang JY, Zhang Y, Li J, Wei L. Intranasal delivery of hypoxia-preconditioned bone marrow-derived mesenchymal stem cells enhanced regenerative effects after intracerebral hemorrhagic stroke in mice. Exp Neurol. 2015;272:78–87. https://doi.org/10.1016/j.expneurol.2015.03.011.
85. Zhang Y, Chen Y, Wu J, Manaenko A, Yang P, Tang J, Fu W, Zhang JH. Activation of dopamine D2 receptor suppresses neuroinflammation through alphaB-crystalline by inhibition of NF-kappaB nuclear translocation in experimental ICH mice model. Stroke. 2015;46(9):2637–46. https://doi.org/10.1161/STROKEAHA.115.009792.
86. Kuric E, Ruscher K. Reduction of rat brain CD8+ T-cells by levodopa/benserazide treatment after experimental stroke. Eur J Neurosci. 2014;40(2):2463–70. https://doi.org/10.1111/ejn.12598.
87. Kuric E, Ruscher K. Reversal of stroke induced lymphocytopenia by levodopa/benserazide treatment. J Neuroimmunol. 2014;269(1–2):94–7. https://doi.org/10.1016/j.jneuroim.2014.02.009.
88. Tsuchiyama R, Sozen T, Manaenko A, Zhang JH, Tang J. The effects of nicotinamide adenine dinucleotide on intracerebral hemorrhage-induced brain injury in mice. Neurol Res. 2009;31(2):179–82. https://doi.org/10.1179/174313209X393609.
89. Ying W, Wei G, Wang D, Wang Q, Tang X, Shi J, Zhang P, Lu H. Intranasal administration with NAD+ profoundly decreases brain injury in a rat model of transient focal ischemia. Front Biosci. 2007;12:2728–34.
90. Belokoskova SG, Dorofeeva SA, Klement'ev BI, Balunov OA. The clinical evaluation of vasopressin in the treatment of aphasias. Zh Nevrol Psikhiatr Im S S Korsakova. 1998;98(7):25–8.
91. Tsikunov SG, Belokoskova SG. Psychophysiological analysis of the influence of vasopressin on speech in patients with post-stroke aphasias. Span J Psychol. 2007;10(1):178–88.
92. Belokoskova SG, Tsikunov SG, Klement'ev BI. Neuropeptide induction of compensatory processes at aphasias. Vestn Ross Akad Med Nauk. 2002;12:28–32.
93. Belokoskova SG, Stepanov II, Tsikunov SG. Agonist of V2 vasopressin receptor reduces depressive disorders in post-stroke patients. Vestn Ross Akad Med Nauk. 2012;4:40–4.
94. Lochhead JJ, Thorne RG. Intranasal delivery of biologics to the central nervous system. Adv Drug Deliv Rev. 2012;64(7):614–28. https://doi.org/10.1016/j.addr.2011.11.002.
95. Banks WA, During MJ, Niehoff ML. Brain uptake of the glucagon-like peptide-1 antagonist exendin(9-39) after intranasal administration. J Pharmacol Exp Ther. 2004;309(2):469–75. https://doi.org/10.1124/jpet.103.063222.
96. Gizurarson S. Anatomical and histological factors affecting intranasal drug and vaccine delivery. Curr Drug Deliv. 2012;9(6):566–82.
97. Selam JL, Slama G. Insulin administration systems. Possibilities and difficulties. Presse Med. 1992;21(33):1575–80.
98. Frey WH 2nd. (WO/1991/007947) Neurologic agents for nasal administration to the brain (priority date 5.12.89). Geneva: World Intellectual Property Organization; 1991.

Chapter 4
Intranasal Drug Delivery After Intracerebral Hemorrhage

Jing Chen-Roetling and Raymond F. Regan

Abstract Intracerebral hemorrhage (ICH) accounts for 10–15% of strokes and is associated with high mortality and disability rates. Studies in animal models suggest a complex pathophysiology mediated by multiple injury cascades that are initiated in the hours after hemorrhage. Effective neuroprotection will therefore likely require rapid delivery of a combination of therapies to perihematomal tissue at risk while minimizing adverse systemic effects. These aims are unlikely to be accomplished by exclusive reliance on intravenous drug administration due to delays in blood-brain barrier penetration and the additive toxicities of multiple agents. In the prehospital and emergency department settings, intranasal drug delivery is the only method currently available that may offer any selective brain targeting. Intranasal administration of recombinant proteins, small molecules and mesenchymal stem cells has improved outcome in both collagenase and blood injection ICH models. These results and the potential utility of intranasal therapies after ICH are reviewed and discussed in this chapter.

Keywords Intracerebral hemorrhage · Neuroprotection · Stroke · Subarachnoid hemorrhage

Spontaneous intracerebral hemorrhage (ICH) is the initial event in 10–15% of strokes. The annual incidence varies from 10 to 30 per 100,000 population, accounting for approximately two million strokes worldwide [1–3]. A number of modifiable ICH risk factors have been identified, including hypertension, alcohol consumption, diabetes, and elevated body mass index, but this information has not yet translated

---

J. Chen-Roetling
Department of Emergency Medicine, Sidney Kimmel Medical College, Thomas Jefferson University, Philadelphia, PA, USA

R. F. Regan (✉)
Department of Emergency Medicine, University of Maryland School of Medicine, Baltimore, MD, USA
e-mail: rregan@som.umaryland.edu

into a reduction in incidence [4]. Prognosis is grim, and despite advances in diagnosis and neurointensive care, has not appreciably improved over the past two decades [1]. The mortality rate at 1 month approaches 50%, and only ~20% of survivors have recovered sufficient function to live independently at this time point [5, 6]. Surgical reduction or removal of the hematoma has been found to benefit only a small subset of patients, and medical management remains largely supportive [7]. Compared with ischemic stroke, fewer pharmacotherapies have progressed to clinical trials, likely a consequence of less intensive preclinical research. To date, none have been found to improve outcome.

4.1 Pathophysiology

Experimental models have delineated a complex series of events that mediates perihematomal cell injury. In the two most commonly used animal models, ICH is produced by stereotactic injection of autologous blood or collagenase into the brain parenchyma. These models emphasize different aspects of ICH, and together provide complementary information that may inform the development of a therapeutic strategy. The primary event of spontaneous ICH is vessel rupture, which immediately delivers blood under arterial pressure to surrounding tissue. This produces tissue deformity and decreases blood flow as microvessels are compressed by tissue pressures that exceed their intralumenal pressures. Classically, all perihematomal injury was attributed to ischemia produced by this mass effect. In support of this hypothesis, rapid and transient inflation of a balloon in the basal ganglia was sufficient per se to produce tissue injury [8]. However, the perihematomal edema characteristic of ICH was not observed in this model, suggesting the participation of additional injury cascades that may be more amenable to therapeutic intervention. Subsequent animal studies indicated that any ischemia produced by the mass effect of the hematoma is at most transient. After injection of a large volume of blood into the dog striatum, perihematomal blood flow was greater than that needed to produce ischemia within 5 h [9]. Perfusion imaging in the acute or subacute phase after clinical ICH was consistent with these experimental observations, since reduction in blood flow was found to be due to decreased metabolic demand rather than to ischemia [10].

Animal studies have demonstrated that secondary injury to perihematomal tissue is characterized by severe blood-brain barrier disruption, edema, migration and activation of inflammatory cells, and cytotoxicity. A considerable and growing body of experimental evidence indicates that this process is initiated at least in part by the release of toxins from the evolving hematoma. Identification of therapies that antagonize the activity of these toxins is a longstanding and active focus of preclinical research. Evidence to date supports the sequential and interactive participation of the following:

4.1.1 Excitotoxicity

Any compressive ischemia is likely associated with rapid release of glutamate and other excitatory neurotransmitters to produce an excitotoxic injury, mediated by excessive activation of NMDA and AMPA receptors. Although the initial insult is rapid and likely not a feasible therapeutic target, it is generally not appreciated that parenchymal blood is a potential source of ongoing excitotoxic stress. Serum has a glutamate concentration of ~40–70 μM [11], increasing towards the upper limit with aging; these concentrations are more than sufficient to kill primary cultured neurons [12]. Even after clearance of serum glutamate, ongoing lysis of erythrocytes may produce continuous excitotoxic stress due to their high micromolar concentrations of glutamate and aspartate [13, 14]. Antagonists of NMDA and AMPA glutamate receptors have been moderately protective alone in ICH models, but are more effective as a component of combinatorial therapy [15–18].

4.1.2 Thrombin

Thrombin is a serine protease that may produce early injury (within 24 h in rodent models) after experimental ICH [19]. Direct injection of activated thrombin into the CNS produces very rapid blood-brain barrier breakdown, edema, and inflammation. Its toxicity may be due in part to interaction with protease-activated receptors (PAR's), particularly PAR-1, which in turn activate Src kinases [20]. The latter phosphorylate and enhance the function of matrix metalloproteinases and NMDA receptors, which in excess may contribute to peri-hematomal edema and cell loss. Thrombin also promotes microglial survival [21], activates the complement cascade [22], and may directly increase oxidative stress by upregulating neuronal NADPH oxidase [23]. Thrombin inhibitors have been protective after experimental ICH [24, 25], but their potent anticoagulant effect will likely contraindicate systemic therapy early after acute ICH.

4.1.3 Hemoglobin and Iron

In the blood injection ICH model, Perls' staining suggested a qualitative increase in iron in perihematomal neurons by 24 h, but a significant increase was not quantified until 3 days [26]. The heme groups of hemoglobin, present at a concentration of ~20 mM in whole blood, are an obvious and frequently-cited source of this iron, since their breakdown by the heme oxygenase (HO) enzymes releases equimolar ferrous iron. In support of the hemoglobin toxicity hypothesis, HO inhibitors reduce edema and decrease injury after blood or hemoglobin infusion into the rodent or pig brain [27, 28]. However, neural cells are initially protected from hemoglobin by the erythrocyte cell membrane. Hemolysis in the hematoma is delayed for 1–3 days in rodent models and in vitro, and perhaps for as long as 1 week in humans [17, 29, 30], although a recent MRI study has challenged the latter assumption [31]. Another, often overlooked source of early iron toxicity is the transferrin-bound iron in plasma, which at physiological concentrations is toxic to cultured neurons [32]. A contribution of transferrin-bound iron to perihematomal cell injury is consistent with the 12 h effective time window for the iron chelator deferoxamine after rodent ICH [33].

Mechanistic interactions likely lower the threshold for toxicity, and contribute to the overall complexity of secondary injury. For example, thrombin upregulates NMDA receptor responses [34], while hemoglobin increases the vulnerability of neurons to excessive AMPA receptor activation [35], potentiating excitotoxicity. Hemin, the oxidized form of heme, is released from extracellular hemoglobin after its spontaneous autoxidation and activates toll-like receptor-4 (Tlr4), synergizing with thrombin to compound the inflammatory response [36]. The multifaceted nature of secondary injury suggests that any therapy that selectively targets a single mechanism is unlikely to be effective in a variable clinical setting, even if protection is observed under highly-controlled laboratory conditions.

4.2 Neuroprotective Considerations

Implementation of a therapeutic strategy targeting the multiple secondary injury cascades identified in experimental models is unlikely to be successful unless it achieves three aims. First, drugs must be delivered rapidly at therapeutic concentrations to perihematomal tissue at risk. Injury pathways are likely to be established within hours, particularly those initiated by thrombin and excitatory amino acids. Second, given the pathophysiological complexity, combinatorial approaches that target multiple parallel cascades will likely be needed. Third, dosing and route of administration must be designed to minimize adverse systemic effects while achieving therapeutic CNS drug concentrations. These aims are unlikely to be accomplished by reliance on the usual methods of emergency drug delivery, due to delays in crossing the blood-brain barrier and the additive toxicities of multiple agents.

4.3 Emergency Drug Administration

Two routes of drug administration, oral and intravenous, are commonly used in emergency settings to treat or prevent acute CNS disease. Both have significant drawbacks for the treatment of hemorrhagic stroke. The oral route is generally impractical, since many acute ICH patients have profound nausea, vomiting, lethargy and swallowing dysfunction. Even if an oral drug could be safely administered and was tolerated, delayed and inconsistent absorption and the first pass hepatic metabolism would decrease the probability of attaining therapeutic serum drug levels in a timely fashion. These limitations are avoided with intravenous drug administration. However, CNS delivery will be impeded by an intact blood-brain barrier, which excludes most hydrophilic and larger lipophilic compounds from the CNS. Delayed bioavailability after systemic drug administration will reduce and perhaps eliminate any neuroprotective benefit, particularly when targeting early injury mechanisms mediated by thrombin and excitotoxicity. Furthermore, the high serum levels needed to obtain therapeutic CNS concentrations increase the likelihood of adverse hemodynamic consequences or off-target toxicity, particularly with rapid administration of multiple agents.

The most reliable method to therapeutically target tissue at risk after ICH is to inject the agent into the adjacent ventricle or directly into the hematoma. These invasive approaches are unavailable in the emergency department and require several hours to mobilize the needed personnel and facilities. While unsuitable for rapid drug administration, intracerebral drug administration may be useful at later time points if it can be accomplished with acceptably low rates of hemorrhagic and infectious complications. However, delivery to the perihematomal region may be delayed and reduced by the slow and restricted diffusion of many drugs through CNS tissue and rapid CSF turnover [37].

Intranasal administration has been used to deliver a wide variety of small molecules and proteins to the CNS, and offers certain advantages over more traditional methods [37]. These include rapid delivery via cranial nerves and perhaps other pathways, reduced systemic exposure, and reduction in adverse effects. In the prehospital and emergency department settings, intranasal drug administration is the only method currently available that may offer any selective brain targeting. While this route is currently underutilized, recent introduction of intranasal naloxone for patients with opioid overdose and benzodiazepines for seizures highlights its utility for rapid delivery of therapeutic drug levels to the CNS [38].

4.4 Intranasal Therapeutics in Animal ICH Models

Investigation of intranasal drug delivery after ICH is currently in its infancy, with only eight animal trials published to date (Table 4.1). These studies have tested the efficacy of acute treatment with four recombinant proteins and two low molecular weight compounds, and delayed treatment with hypoxia-preconditioned bone marrow-derived mesenchymal stem cells to enhance recovery. All but one of these studies are from the same research group.

Table 4.1 Published studies testing intranasal recombinant protein, small molecule and stem cell therapies in rodent blood injection or collagenase ICH models

Treatment	Species	Model	Result	Reference
IGF-1	Rat Pup	Collagenase	+	Lekic et al. [39]
Osteopontin	Rat	Collagenase	+	Malaguit et al. [40]
Adropin	Mouse	Collagenase	+	Yu et al. [41]
Gas6	Mouse	Blood Inj	+	Tong et al. [42]
Osteopontin	Rat	Collagenase	+	Gong et al. [43]
NAD	Mouse	Collagenase	−	Tsuchiyama et al. [44]
Quinpirole	Mouse	Both	+	Zhang et al. [45]
Mesenchymal stem cells	Mouse	Collagenase	+	Sun et al. [46]

4.4.1 Recombinant Proteins

Direct nose to brain protein transport was first reported over two decades ago [47]. In the absence of selective transport mechanisms, access of hydrophilic proteins to the CNS after parenteral administration is severely limited by an intact blood-brain barrier. Attaining therapeutic levels of most proteins by intravenous administration, if ever feasible, will be delayed by the several hours to days needed for barrier disruption. This interval will often exceed the therapeutic time window for a given injury mechanism. Intranasal administration of recombinant neurotrophic factors and other regulatory proteins may be their most feasible delivery route, and may be facilitated by their potency, with EC_{50} values at low nanomolar concentrations. Efficacy in an ICH model was first reported by Lekic et al. for insulin-like growth factor 1 (IGF-1), a neurotrophin with pleiotropic neuroprotective effects [39]. When administered acutely after collagenase-induced germinal matrix hemorrhage in neonatal rats, it reduced blood-brain barrier disruption, edema, and neurological deficits at 3 days and cognitive deficits at 3 weeks. Similar results were reported in rodents treated at 1 h after hemorrhage induction with intranasal recombinant adropin [41] or growth arrest-specific 6 (Gas6) [42] in adult striatal ICH models, and osteopontin in adult and neonatal ICH models [40, 43].

4.4.2 Small Molecules

The detailed preclinical investigation and recent clinical introduction of intranasal naloxone and benzodiazepines demonstrate the utility of this route for delivery of pharmacotherapies in emergency settings. However, despite a plethora of drug studies published in experimental ICH models, only two have utilized this approach to date. Zhang et al. reported that intranasal administration of the dopamine D2 receptor agonist quinpirole at 1 h after striatal injection of autologous blood in a mouse model reduced edema and neurologicial deficits [45]. Intraperitoneal injection had a similar effect, which is not surprising since quinpirole readily crosses the blood-brain barrier when given parenterally [48]. Quinpirole blood and brain levels were not quantified, so it is not clear if the efficacy of intranasal administration was due to direct nose to brain delivery or to vascular uptake. In the only other small molecule study, Tsuchiyama et al. reported nicotinamide adenine dinucleotide (NAD^+) was successfully delivered to the mouse cerebral hemispheres by intranasal administration but had no effect on edema or brain water content after collagenase-induced ICH [44].

4.4.3 Stem Cells

Perhaps the most surprising observation in nose to brain delivery research is that intranasal stem cells enter the CNS in sufficient numbers to produce a therapeutic benefit [49]. In rodents, this transit occurs across the olfactory epithelium to the subarachnoid space at the cribiform plate, with subsequent migration to the site of injury [50]. Consistent with observations in rodent models of ischemic stroke, subarachnoid hemorrhage and Parkinson's disease [38, 51], Sun et al. reported promising results in a mouse collagenase ICH model [46]. Mesenchymal stem cells were harvested from rat bone marrow, cultured, subjected to hypoxic preconditioning, and administered intranasally at 3 and 7 days after ICH. These cells were observed in perihematomal tissue and were associated with significant increases in tissue neurotrophin levels, neurogenesis, and functional recovery. Intranasal stem cell delivery may therefore provide a novel approach to reparative therapy with a very long therapeutic window, while avoiding the adverse hemorrhagic, inflammatory and infectious complications of surgical stem cell transplantation.

These key trials clearly support the utility of the intranasal route in rodent models, and provide a compelling rationale for further investigation. However, two key issues remain to be addressed. First, the relevance of rodent models to clinical nose-to-brain drug delivery research should be addressed, due to the considerable anatomical differences of rodents and humans. The olfactory epithelium, the primary site of nose-to-brain delivery, lines 50% of the rat nasal cavity but only 3% in humans, likely impacting uptake and CNS bioavailability [52]. Identification of appropriate large animal models and detailed drug trials using these models are prerequisites for any of clinical testing. Second, it is essential to determine by direct comparative trials if the intranasal route provides any benefit over intravenous administration. In addition to quantifying efficacy, blood, CSF and CNS tissue concentrations should be determined over time after both intranasal and intravenous administration. The CNS targeting ratio, defined as the ratio of brain to blood drug concentration after intranasal administration divided by this ratio after intravenous administration [53], should be calculated. Low ratios likely indicate vascular uptake via the nasal respiratory epithelium, which is rapid and efficient for many lipophilic drugs. However, drugs with low CNS targeting ratios are poor candidates for intranasal delivery after ICH, since vascular access is rarely unavailable for these patients in a modern Emergency Department, and intranasal administration will not attenuate any adverse effects of intravenous administration.

4.5 Pitfalls of Systemic Therapy: The Case of Deferoxamine

Deferoxamine (DFO) is a metal chelator that was initially identified as a siderophore produced by Streptomyces pilosus, and was subsequently synthesized for pharmaceutical use. It was the focus of a recent clinical trial that highlights the challenges of treating ICH with a pharmaceutical compound that is not targeted to the CNS. DFO has very high affinity for ferric iron (stability constant $10^{31)}$ [54]) and as a hexadentate chelator it efficiently binds all six iron coordination sites, rendering it essentially unreactive. In aqueous solutions at neutral pH, DFO is a positively charged amine that penetrates cell membranes poorly. However, it is taken up by endocytosis and is fortuitously concentrated in lysosomes [55], a compartment containing most of the cellular labile iron that may catalyze oxidative reactions [56]. Its protective effect against the oxidative toxicity of hemoglobin was first demonstrated in vitro [57, 58], providing the rationale for testing its effect in ICH models. Using a rat blood injection ICH model, Nakamura et al. reported that bolus injections of 100 mg/kg i.p. twice daily reduced edema, behavioral deficits and oxidative injury markers when begun at 2–6 h after striatal blood injection [59]. Similar results were subsequently observed in aged rats and piglets [33, 60], but efficacy in the collagenase ICH model was inconsistent [61, 62].

Deferoxamine has a number of limitations as a systemic therapeutic agent that were recognized during decades of clinical use to treat acute iron poisoning, transfusion-related iron overload, and hemochromatosis. When given by bolus intravenous (i.v.) injection, it is rapidly eliminated via renal excretion, and has a plasma half life of only 5–15 min [63, 64]. Intravenous administration is occasionally complicated by hypotension, and for that reason it is usually administered to patients by continuous subcutaneous or intramuscular (i.m.) infusion. As a polar hydrophilic compound with a relatively high molecular weight, it penetrates the intact blood-brain barrier poorly [55, 64], which may account in part for the high i.m. or i.p. doses required for neuroprotection after ICH in animal models (100–200 mg/kg/day), which approach the LD_{50} for mice (300 mg/kg) [65]. With prolonged use, the toxicity of DFO is likely due to its ability to sequester iron and other metals in a form that is inaccessible to cells, resulting in iron starvation and metalloenzyme inhibition [66, 67]. It is tolerated at doses of up to 50 mg/kg/day in patients with systemic iron overload, but the dose must be rapidly reduced as iron concentrations decline [68]. ICH patients usually will not have systemic iron overload, and so toxic effects may be more rapidly manifested. In the initial multicenter trial (High-Dose Deferoxamine in Intracerebral Hemorrhage, HI-DEF [69]), patients receiving 62 mg/kg/day (maximum of 6 g/day) for 5 days had an increased incidence of ARDS, an known DFO adverse effect first reported in 1992 [70], leading to trial suspension. A second trial (iDEF Trial) using a lower dose (32 mg/kg/day for 3 days) is currently active.

The adverse hemodynamic, pharmacokinetic, and toxicity profile of intravenous DFO provides a strong rationale for more specific targeting to the CNS to treat hemorrhagic stroke. Studies to date indicate that intranasal DFO rapidly produces therapeutic DFO concentrations throughout the brain with relatively low blood levels, at least in rodents. Hanson et al. [53] reported that 30 min after intranasal administration of a high DFO dose (6 mg) to uninjured rats, brain tissue concentrations ranged from 0.9 to 18.5 µM. The highest concentration was in the olfactory bulb and the lowest was in the cerebellum. Cortical DFO levels ranged from 1.5 to 3.4 µM, and the striatal concentration was 1.7 µM; blood concentration at the same time point was 1.6 µM. After intravenous injection of the same dose, brain concentrations were very low (0.14–0.54 µM), with the highest concentration again in the olfactory bulb. Cortical DFO levels ranged from 0.25 to 0.35 µM, the striatal concentration was only 0.14 µM and blood concentration was 43.7 µM. The targeting ratios, as defined above, were 173–271 in the cortex and 332 in the striatum. In the same study, intranasal DFO at this dose reduced infarct volume after middle cerebral artery occlusion, without clinically significant alterations in blood pressure or heart rate.

The high intranasal DFO dose used in the Hanson et al. study may not be feasible for the multiple doses required after ICH due to its adverse irritant effects. More recent experiments indicate that the CNS bioavailability of intranasal DFO can be markedly enhanced by encapsulating it in solid microparticles constructed with spherical chitosan chloride or methyl-β-cyclodextrins [64]. These microparticles are mucoadhesive, thereby increasing retention time in the nasal cavity. When hydrated by local secretions, they swell and release their DFO content. Administration of only 200 µg encapsulated DFO to healthy rats, about 3% of the dose used in the Hanson et al. study, produced CSF DFO concentrations of 3.83 ± 0.68 µg/mL (6.83 ± 1.21 µM) for chitosan microparticles and 14.37 ± 1.69 µg/mL (25.61 ± 3.01 µM) for methyl-β-cyclodextrin microparticles at 30 min. In contrast, peak plasma concentrations were only 0.19 ± 0.04 µg/mL (0.33 ± 0.07 µM) and 0.36 ± 0.06 µg/mL (0.64 ± 0.11 µM), respectively, indicating minimal transport of encapsulated nasal DFO into the systemic circulation. At this low dose an aqueous solution of DFO did not produce significant CSF levels with either intravenous or intranasal administration. These results demonstrate that the CNS targeting and bioavailability of a nasally administered drug is can vary significantly with the delivery vehicle.

4.6 Conclusions

An effective pharmacotherapy to attenuate secondary injury cascades after ICH will likely require rapid delivery of multiple agents to CNS tissue at risk. The need to bypass an intact blood-brain barrier while minimizing the additive toxicities of multiple agents creates a significant challenge, particularly during the early hours after hemorrhage. Invasive approaches are largely unavailable in the prehospital or emergency department setting, and intranasal administration may be the only available option to selectively target small molecule or recombinant protein therapies to the brain. Published results to date provide a compelling rationale for further investigation of intranasal drug delivery in appropriate ICH animal models. Future studies should focus on pharmacokinetics, direct comparison of efficacy with systemic and intranasal administration, and optimal formulations to maximize brain uptake while minimizing blood levels of the compound of interest.

References

1. Rincon F, Mayer SA. The epidemiology of intracerebral hemorrhage in the United States from 1979 to 2008. Neurocrit Care. 2013;19:95–102.
2. Qureshi AI, Mendelow AD, Hanley DF. Intracerebral haemorrhage. Lancet. 2009;373:1632–44.
3. Taylor TN, Davis PH, Torner JC, Holmes J, Meyer JW, Jacobson MF. Lifetime cost of stroke in the United States. Stroke. 1996;27:1459–66.
4. Poon MT, Bell SM, Al-Shahi Salman R. Epidemiology of intracerebral haemorrhage. Front Neurol Neurosci. 2015;37:1–12.
5. Broderick JP, Adams HP Jr, Barsan W, Feinberg W, Feldmann E, Grotta J, Kase C, Krieger D, Mayberg M, Tilley B, et al. Guidelines for the management of spontaneous intracerebral hemorrhage: a statement for healthcare professionals from a special writing group of the Stroke Council, American Heart Association. Stroke. 1999;30:905–15.
6. Sacco S, Marini C, Toni D, Olivieri L, Carolei A. Incidence and 10-year survival of intracerebral hemorrhage in a population-based registry. Stroke. 2009;40:394–9.
7. Hemphill JC 3rd, Greenberg SM, Anderson CS, Becker K, Bendok BR, Cushman M, Fung GL, Goldstein JN, Macdonald RL, Mitchell PH, et al. Guidelines for the management of spontaneous intracerebral hemorrhage: a guideline for healthcare professionals from the American Heart Association/American Stroke Association. Stroke. 2015;46:2032–60.
8. Sinar EJ, Mendelow AD, Graham DI, Teasdale GM. Experimental intracerebral hemorrhage: effects of a temporary mass lesion. J Neurosurg. 1987;66:568–76.
9. Qureshi AI, Wilson DA, Hanley DF, Traystman RJ. No evidence for an ischemic penumbra in massive experimental intracerebral hemorrhage. Neurology. 1999;52:266–72.
10. Zazulia AR, Diringer MN, Videen TO, Adams RE, Yundt K, Aiyagari V, Grubb RL Jr, Powers WJ. Hypoperfusion without ischemia surrounding acute intracerebral hemorrhage. J Cereb Blood Flow Metab. 2001;21:804–10.
11. Pitkanen HT, Oja SS, Kemppainen K, Seppa JM, Mero AA. Serum amino acid concentrations in aging men and women. Amino Acids. 2003;24:413–21.
12. Schramm M, Eimerl S, Costa E. Serum and depolarizing agents cause acute neurotoxicity in cultured cerebellar granule cells: role of the glutamate receptor responsive to n-methyl-d-aspartate. Proc Natl Acad Sci U S A. 1990;87:1193–7.
13. D'Eufemia P, Finocchiaro R, Lendvai D, Celli M, Viozzi L, Troiani P, Turri E, Giardini O. Erythrocyte and plasma levels of glutamate and aspartate in children affected by migraine. Cephalalgia. 1997;17:652–7.
14. Divino Filho JC, Hazel SJ, Furst P, Bergstrom J, Hall K. Glutamate concentration in plasma, erythrocyte and muscle in relation to plasma levels of insulin-like growth factor (igf)-i, igf binding protein-1 and insulin in patients on haemodialysis. J Endocrinol. 1998;156:519–27.
15. Ardizzone TD, Lu A, Wagner KR, Tang Y, Ran R, Sharp FR. Glutamate receptor blockade attenuates glucose hypermetabolism in perihematomal brain after experimental intracerebral hemorrhage in rat. Stroke. 2004;35:2587–91.
16. Sinn DI, Lee ST, Chu K, Jung KH, Song EC, Kim JM, Park DK, Kim M, Roh JK. Combined neuroprotective effects of celecoxib and memantine in experimental intracerebral hemorrhage. Neurosci Lett. 2007;411:238–42.

17. Jaremko KM, Chen-Roetling J, Chen L, Regan RF. Accelerated hemolysis and neurotoxicity in neuron-glia-blood clot co-cultures. J Neurochem. 2010;114:1063–73.
18. Terai K, Suzuki M, Sasamata M, Yatsugi S, Yamaguchi T, Miyata K. Effect of ampa receptor antagonist ym872 on cerebral hematoma size and neurological recovery in the intracerebral hemorrhage rat model. Eur J Pharmacol. 2003;467:95–101.
19. Hua Y, Keep RF, Hoff JT, Xi G. Brain injury after intracerebral hemorrhage: the role of thrombin and iron. Stroke. 2007;38:759–62.
20. Ardizzone TD, Zhan X, Ander BP, Sharp FR. Src kinase inhibition improves acute outcomes after experimental intracerebral hemorrhage. Stroke. 2007;38:1621–5.
21. Ohnishi M, Katsuki H, Izumi Y, Kume T, Takada-Takatori Y, Akaike A. Mitogen-activated protein kinases support survival of activated microglia that mediate thrombin-induced striatal injury in organotypic slice culture. J Neurosci Res. 2010;88:2155–64.
22. Gong Y, Xi GH, Keep RF, Hoff JT, Hua Y. Complement inhibition attenuates brain edema and neurological deficits induced by thrombin. Acta Neurochir Suppl. 2005;95:389–92.
23. Park KW, Jin BK. Thrombin-induced oxidative stress contributes to the death of hippocampal neurons: role of neuronal nadph oxidase. J Neurosci Res. 2008;86:1053–63.
24. Lee KR, Colon GP, Betz AL, Keep RF, Kim S, Hoff JT. Edema from intracerebral hemorrhage: the role of thrombin. J Neurosurg. 1996;84:91–6.
25. Kitaoka T, Hua Y, Xi G, Hoff JT, Keep RF. Delayed argatroban treatment reduces edema in a rat model of intracerebral hemorrhage. Stroke. 2002;33:3012–8.
26. Wu J, Hua Y, Keep RF, Nakemura T, Hoff JT, Xi G. Iron and iron-handling proteins in the brain after intracerebral hemorrhage. Stroke. 2003;34:2964–9.
27. Wagner KR, Hua Y, de Courten-Myers GM, Broderick JP, Nishimura RN, Lu SY, Dwyer BE. Tin-mesoporphyrin, a potent heme oxygenase inhibitor, for treatment of intracerebral hemorrhage: in vivo and in vitro studies. Cell Mol Biol. 2000;46:597–608.
28. Huang FP, Xi G, Keep RF, Hua Y, Nemoianu A, Hoff JT. Brain edema after experimental intracerebral hemorrhage: role of hemoglobin degradation products. J Neurosurg. 2002;96:287–93.
29. Bradley WG Jr. MR appearance of hemorrhage in the brain. Radiology. 1993;189:15–26.
30. Wagner KR, Dwyer BE. Hematoma removal, heme, and heme oxygenase following hemorrhagic stroke. Ann N Y Acad Sci. 2004;1012:237–51.
31. Liu R, Li H, Hua Y, Keep RF, Xiao J, Xi G, Huang Y. Early hemolysis within human intracerebral hematomas: an MRI study. Transl Stroke Res. 2018. https://doi.org/10.1007/s12975-018-0630-2.
32. Chen-Roetling J, Liu W, Regan RF. Iron accumulation and neurotoxicity in cortical cultures treated with holotransferrin. Free Radic Biol Med. 2011;51:1966–74.
33. Okauchi M, Hua Y, Keep RF, Morgenstern LB, Schallert T, Xi G. Deferoxamine treatment for intracerebral hemorrhage in aged rats. Therapeutic time window and optimal duration. Stroke. 2010;41:375–82.
34. Gingrich MB, Junge CE, Lyuboslavsky P, Traynelis SF. Potentiation of nmda receptor function by the serine protease thrombin. J Neurosci. 2000;20:4582–95.
35. Regan RF, Panter SS. Hemoglobin potentiates excitotoxic injury in cortical cell culture. J Neurotrauma. 1996;13:223–31.
36. Lin S, Yin Q, Zhong Q, Lv FL, Zhou Y, Li JQ, Wang JZ, Su BY, Yang QW. Heme activates TLR4-mediated inflammatory injury via MyD88/TRIF signaling pathway in intracerebral hemorrhage. J Neuroinflammation. 2012;9:46.
37. Dhuria SV, Hanson LR, Frey WH 2nd. Intranasal delivery to the central nervous system: mechanisms and experimental considerations. J Pharm Sci. 2010;99(4):1654–73.
38. Chapman CD, Frey WH 2nd, Craft S, Danielyan L, Hallschmid M, Schioth HB, Benedict C. Intranasal treatment of central nervous system dysfunction in humans. Pharm Res. 2013;30:2475–84.
39. Lekic T, Flores J, Klebe D, Doycheva D, Rolland WB, Tang J, Zhang JH. Intranasal igf-1 reduced rat pup germinal matrix hemorrhage. Acta Neurochir Suppl. 2016;121:209–12.
40. Malaguit J, Casel D, Dixon B, Doycheva D, Tang J, Zhang JH, Lekic T. Intranasal osteopontin for rodent germinal matrix hemorrhage. Acta Neurochir Suppl. 2016;121:217–20.

41. Yu L, Lu Z, Burchell S, Nowrangi D, Manaenko A, Li X, Xu Y, Xu N, Tang J, Dai H, et al. Adropin preserves the blood-brain barrier through a notch1/hes1 pathway after intracerebral hemorrhage in mice. J Neurochem. 2017;143:750–60.
42. Tong LS, Shao AW, Ou YB, Guo ZN, Manaenko A, Dixon BJ, Tang J, Lou M, Zhang JH. Recombinant Gas6 augments Axl and facilitates immune restoration in an intracerebral hemorrhage mouse model. J Cereb Blood Flow Metab. 2017;37:1971–81.
43. Gong L, Manaenko A, Fan R, Huang L, Enkhjargal B, McBride D, Ding Y, Tang J, Xiao X, Zhang JH. Osteopontin attenuates inflammation via JAK2/STAT1 pathway in hyperglycemic rats after intracerebral hemorrhage. Neuropharmacology. 2018;138:160–9.
44. Tsuchiyama R, Sozen T, Manaenko A, Zhang JH, Tang J. The effects of nicotinamide adenine dinucleotide on intracerebral hemorrhage-induced brain injury in mice. Neurol Res. 2009;31:179–82.
45. Zhang Y, Chen Y, Wu J, Manaenko A, Yang P, Tang J, Fu W, Zhang JH. Activation of dopamine D2 receptor suppresses neuroinflammation through αB-crystalline by inhibition of NF-κB nuclear translocation in experimental ICH mice model. Stroke. 2015;46:2637–46.
46. Sun J, Wei ZZ, Gu X, Zhang JY, Zhang Y, Li J, Wei L. Intranasal delivery of hypoxia-preconditioned bone marrow-derived mesenchymal stem cells enhanced regenerative effects after intracerebral hemorrhagic stroke in mice. Exp Neurol. 2015;272:78–87.
47. Thorne RG, Emory CR, Ala TA, Frey WH 2nd. Quantitative analysis of the olfactory pathway for drug delivery to the brain. Brain Res. 1995;692:278–82.
48. Charbit AR, Akerman S, Goadsby PJ. Comparison of the effects of central and peripheral dopamine receptor activation on evoked firing in the trigeminocervical complex. J Pharmacol Exp Ther. 2009;331:752–63.
49. Danielyan L, Schafer R, von Ameln-Mayerhofer A, Buadze M, Geisler J, Klopfer T, Burkhardt U, Proksch B, Verleysdonk S, Ayturan M, et al. Intranasal delivery of cells to the brain. Eur J Cell Biol. 2009;88:315–24.
50. Galeano C, Qiu Z, Mishra A, Farnsworth SL, Hemmi JJ, Moreira A, Edenhoffer P, Hornsby PJ. The route by which intranasally delivered stem cells enter the central nervous system. Cell Transplant. 2018;27:501–14.
51. Nijboer CH, Kooijman E, van Velthoven CT, van Tilborg E, Tiebosch IA, Eijkelkamp N, Dijkhuizen RM, Kesecioglu J, Heijnen CJ. Intranasal stem cell treatment as a novel therapy for subarachnoid hemorrhage. Stem Cells Dev. 2018;27:313–25.
52. Illum L. Is nose-to-brain transport of drugs in man a reality? J Pharm Pharmacol. 2004;56:3–17.
53. Hanson LR, Roeytenberg A, Martinez PM, Coppes VG, Sweet DC, Rao RJ, Marti DL, Hoekman JD, Matthews RB, Frey WH 2nd, et al. Intranasal deferoxamine provides increased brain exposure and significant protection in rat ischemic stroke. J Pharmacol Exp Ther. 2009;330:679–86.
54. Halliwell B, Gutteridge JMC. Free radicals in biology and medicine. 3rd ed. Oxford: Oxford University Press; 1999. p. 936.
55. Cable H, Lloyd JB. Cellular uptake and release of two contrasting iron chelators. J Pharm Pharmacol. 1999;51:131–4.
56. Persson HL, Richardson DR. Iron-binding drugs targeted to lysosomes: a potential strategy to treat inflammatory lung disorders. Expert Opin Investig Drugs. 2005;14:997–1008.
57. Sadrzadeh SMH, Graf E, Panter SS, Hallaway PE, Eaton JW. Hemoglobin: a biologic Fenton reagent. J Biol Chem. 1984;259:14354–6.
58. Regan RF, Panter SS. Neurotoxicity of hemoglobin in cortical cell culture. Neurosci Lett. 1993;153:219–22.
59. Nakamura T, Keep RF, Hua Y, Schallert T, Hoff JT, Xi G. Deferoxamine-induced attenuation of brain edema and neurological deficits in a rat model of intracerebral hemorrhage. J Neurosurg. 2004;100:672–8.
60. Gu Y, Hua Y, Keep RF, Morgenstern LB, Xi G. Deferoxamine reduces intracerebral hematoma-induced iron accumulation and neuronal death in piglets. Stroke. 2009;40:2241–3.
61. Warkentin LM, Auriat AM, Wowk S, Colbourne F. Failure of deferoxamine, an iron chelator, to improve outcome after collagenase-induced intracerebral hemorrhage in rats. Brain Res. 2010;1309:95–103.

62. Wu H, Wu T, Xu X, Wang J. Iron toxicity in mice with collagenase-induced intracerebral hemorrhage. J Cereb Blood Flow Metab. 2011;31:1243–50.
63. Dragsten PR, Hallaway PE, Hanson GJ, Berger AE, Bernard B, Hedlund BE. First human studies with a high-molecular-weight iron chelator. J Lab Clin Med. 2000;135:57–65.
64. Rassu G, Soddu E, Cossu M, Brundu A, Cerri G, Marchetti N, Ferraro L, Regan RF, Giunchedi P, Gavini E, et al. Solid microparticles based on chitosan or methyl-beta-cyclodextrin: a first formulative approach to increase the nose-to-brain transport of deferoxamine mesylate. J Control Release. 2015;201:68–77.
65. Persson HL, Yu Z, Tirosh O, Eaton JW, Brunk UT. Prevention of oxidant-induced cell death by lysosomotropic iron chelators. Free Radic Biol Med. 2003;34:1295–305.
66. Porter JB, Huehns ER. The toxic effects of desferrioxamine. Baillieres Clin Haematol. 1989;2:459–74.
67. Chaston TB, Richardson DR. Iron chelators for the treatment of iron overload disease: relationship between structure, redox activity, and toxicity. Am J Hematol. 2003;73:200–10.
68. Kushner JP, Porter JP, Olivieri NF. Secondary iron overload. Hematology. 2001;2001:47–61.
69. Yeatts SD, Palesch YY, Moy CS, Selim M. High dose deferoxamine in intracerebral hemorrhage (hi-def) trial: rationale, design, and methods. Neurocrit Care. 2013;19:257–66.
70. Tenenbein M, Kowalski S, Sienko A, Bowden DH, Adamson IY. Pulmonary toxic effects of continuous desferrioxamine administration in acute iron poisoning. Lancet. 1992;339:699–701.

Chapter 5
Intranasal Treatment in Subarachnoid Hemorrhage

Basak Caner

Abstract Aneurysmal subarachnoid hemorrhage is a devastating disease because of the initial mortality rate and delayed cerebral iscemia which can be caused by several pathological mechanisms not fully understood yet. To date there is no specific treatment for the delayed cerebral iscemia. One of the most important challenges in subarachnoid hemorrhage research area is the blood brain barrier. Intranasal treatment is a promising way of treatment to bypass the blood brain barrier. In this chapter intranasal treatment options in subarachnoid hemorrhage will be discussed.

Keywords Subarachnoid hemorrhage · Vasospasm · Intranasal delivery · Early brain injury

Aneurysmal subarachnoid hemorrhage (aSAH) accounts for <5% of all strokes, but it is a devastating disease with an initial 30% mortality and rebleeding risk [1]. The main problem in managing aSAH remains delayed ischemic injury which was attributed to vasospasm for long years. But in the light of recent studies in animals as well as in humans, it has come out that treating vasospasm is not treating the patients outcome, therefore researchers have been seeking for new concepts in aSAH studies [2].

As a new emerging concept early or acute brain injury has taken the attention of aSAH reseachers. As a result of increased intracranial pressure and decreased cerebral blood flow after the initial bleeding, global ischemia initiates a cascade of pathological changes that occur before the onset of delayed vasospasm. These pathological changes promote blood brain barrier disruption (BBB) and increase neuronal cell death, through various mechanisms like inflammation, molecular alterations, microthrombosis and cortical depolarization [1].

B. Caner (✉)
Department of Neurosurgery, Medeniyet University Goztepe Education and Research Hospital, Istanbul, Turkey

Although numerous agents can prevent arterial narrowing and or block the excitatory cascade of events leading to ischemic neuronal death in experimental conditions, attributed whether to vasospasm, delayed ischemic injury or early brain injury; there is still no proven treatment or pharmacologic agent that has been shown conclusively to improve the outcome in clinical practice.

To date the main therapeutic interventions used are limited to manipulation of systemic blood pressure, alteration of blood volume or viscosity and control of arterial carbon dioxide tension [3, 4].

Given the fact that there is no proven treatment strategy in aSAH; the major handicap in the preclinical and clinical studies is still the blood brain barrier, like all the other CNS pathologies. Blood brain barrier as the physiologically protector of the brain with its unique nature of selective permeability presents the key challenge to pharmaceutical treatment of the central nervous system disorders [5].

Many drugs discovered for CNS disorders fail to enter the market because of their inability to cross the BBB. Intraventricular injection can deliver drugs directly to the brain; however it is highly invasive and not realistic for clinical applications.

To overwhelm the BBB problem several approaches like physiological transport mechanism, adsorptive-mediated transcytosis, active efflux transport, carrier mediated transport, receptor mediated transport of drugs have been investigated; intranasal delivery is one of the alternate strategies [6].

There are several advantages of administering drugs via the nasal route. The nasal cavity serves as a direct route for medications that can easily cross mucous membranes because of its rich vascular plexus. Moreover the direct absorption into the blood stream bypasses the gastrointestinal tract and hepatic first pass metabolism therefore allowing more drug to be bioavailable than administered orally [7].

Nasal delivery of drugs provides rapid delivery of molecules to the CNS via bulk flow along olfactory and trigeminal perivascular channels and additional slower delivery via olfactory bulb axonal transport. In the upper nasal passage the dendritic processes of the olfactory neurons are directly exposed. Therefore the uptake of the drugs happen either by endocytosis and transport by the olfactory nerves or by extracellular flow through intercellular clefts [8].

In many studies where the researchers compared the rates of absorption and plasma concentration the rates were similar to intravenous administration and even better than subcutaneous and intramuscular routes [7].

For a variety of growth factors, hormones, neuropeptides and therapeutics including insulin, oxytocin, orexin and even stem cells intranasal delivery is emerging as an efficient method of administration in CNS disorders [9]. Other CNS disorders will be discussed in other chapters of this book.

5.1 Recombinant Osteopontin in Subarachnoid Hemorrhage

One of the mostly studied molecules intranasally administered in aSAH is recombinant Osteopontin (rOPN). Osteopontin is an extracellular matrix protein, which by interacting with its cell surface integrin receptors plays a big role in cell proliferation and in reduction of apoptotic cell death. Intracerebroventricular administration of rOPN after experimental subarachnoid hemorrhage decreased brain edema and attenuated cerebral vasospasm [10].

Taking into consideration that the intracerebroventricular administration is not feasible in clinical treatment strategies rOPN was administered intranasally in an experimental rat subarachnoid hemorrhage model. rOPN was administered 30 min after subarachnoid hemorrhage over a period of 20 min and was detected within the CSF at 3 h after treatment. It has been shown that rOPN improved functional outcome and reduced brain edema and neuronal apoptotic cell death. In this study rOPN exerts its effects over FAK activated PI3K-Akt pathway [11].

The duration between drug administration and its detection in the brain varies based on the molecular structure of the compound and also on the species of different animal species. In one study where rOPN was administered in a mouse stroke model rOPN has been detected after 2 h of administration [12], while in the subarachnoid hemorrhage rat model rOPN was detected after 3 h.

In another study where rOPN was administered inranasally after 3 h of subarachnoid hemorrhage, it prevented the changes of α-smooth muscle actin (a marker of vascular smooth muscle cells) and significantly alleviated neurobehavioral dysfunction, increased the cross-sectional area and lumen diameter of the cerebral arteries, reduced the brain water content and brain swelling, and improved the wall thickness of cerebral arteries.

To gain more insight in the results of rOPN in this study the concept of vascular neural network has to be explained. Vascular neural network emerged as a new concept in subarachnoid hemorrhage studies to redefine the vascular pathophysiology for subarachnoid induced vasospasm and early/delayed brain injury. It is consisted of five compounds as large artery moderate vasospasm, small arterial vasospasm or dilatation, capillary occlusion and compression, venous vasospasm, compression and thrombus formation and large vein vasospasm [13].

This network includes vascular smooth muscle cells, which typically switch from contractile to synthetic type and functionally from contraction to repair and migration after injury, resulting in decreasing autoregulatory capacity and regional blood flow to enhance brain swelling and brain edema. Intranasal rOPN prevented this smooth muscle phenotypic transformation [14].

Since the high cost of rOPN treatment in clinical translation researchers seek for cost-effective treatment methods including osteopontin. It has been shown that Vitamin D upregulates endogenous OPN expression [15]. Pre and after SAH intranasal administration of VitD3 significantly increased neurological function and decreased the brain edema formation, and was associated with upregulation of endogenous OPN-A and -C isomers. Thus intranasal administration would be a cost-effective way to upregulate OPN in SAH patients [16].

5.2 Intranasal Administration of Recombinant Netrin-1

Netrin-1(NTN-1) is a laminin-related protein highly expressed in neurons and has been regarded as an anti-apoptotic molecule in various tissues. Xie et al. found that the level of endogenous NTN-1 is increasing in the left cerebral cortex after subarachnoid hemorrhage. Moreover administration of exogenous rNTN-1 intranasally 1 h after experimental SAH significantly improved neurological functions and attenuated neuronal apoptosis concomitant with upregulating APPL-1, pAkt and BCL2 expressions and downregulating the apoptotic marker Cc-3 expression [17].

5.3 Intranasal Stem Cell Treatment in aSAH

Mesencymal stem cell (MSC) treatment showed promising effects after ischemic brain injury models like stroke and neonatal hypoxia-ischemia in improving functional outcome and decreasing brain lesion volume [18, 19].

Intravenous and intraarterial administration of stem cells resulted in high accumulation of the cells in peripheral organs where they were not needed [18, 20, 21].

In a recent study the authors have studied the effects of intranasal administered bone marrow-derived MSCs in a rat endovascular perforation model of subarachnoid hemorrhage [22].

They administered MSCs in awake animals 6 days after SAH 30 min after two doses of Hyaluronidase administration. Hyaluronidase degrades hyaluronic acid, thereby increasing tissue permeability which is necessary for administration of larger particles like cells.

Intranasal administration improved the sensorimotor function and decreased the lesion size. Moreover neuroinflammation associated with SAH and astrocyte and microglia/macrophage activation was significantly reduced. Finally the depression like state of rats after SAH was reversed after intranasal MSC administration.

5.4 Promising Progress in Intranasal Treatment Research in aSAH

5.4.1 Nimodipine-Loaded Lipopluronics Micelles

To date nimodipine is known as the only FDA-approved drug for treating subarachnoid hemorrhage induced vasospasm. But in the latest evidence-based guidelines for subarachnoid hemorrhage by American Heart Association/American Stroke Association, it is especially pointed out that oral nimodipine has been shown to improve neurological outcomes but not cerebral vasospasm and the value of other calcium antagonists, whether administered orally or intravenously remains uncertain [3].

Nimodipine is a calcium channel blocker, which has been shown to dilate cerebral arterioles and increase cerebral blood flow. The guidelines recommend the administration of 60 mg oral Nimodipine every 6 h. But as a Class II drug according to the Biopharmaceutical Classification System nimodipine suffers from limited oral bioavailability and extensive first pass metabolism in the liver. On the other hand intravenous administration of NM has several problems like hypotension, bradycardia and arrhythmias. In a recent study the authors were able to create nimodipine-loaded lipo-pluronic micelles. Micelles are spherical colloidal systems consisting of a hydrophobic core and hydrophilic surface [6]. Pluronic micelles have a core-shell structure with hydrophobic core acting as solubilization depot for non-polar drugs and hydrophilic corona preventing aggregation, protein absorption and recognition by reticulo-endothelial system which leads to longer blood circulation time. The evaluation of intranasal delivery showed that NM brain peak concentration after IN administration was significantly higher than the peak concentration following iv administration. Furtherfore the blood/brain ratios were significantly higher at all sampling intervals after IN administration [23]. It means that in nano-size prepared NM-loaded LPM can be a promising replacement for the NM iv injection resulting in high brain concentration without having any systemic side-effects.

5.4.2 Erythropoietin (EPO) for SAH

One of the most studied treatment strategies concerning neuroprotection in stroke is EPO. Human EPO is a glycoprotein growth factor that acts as the main regulator of erythropoiesis. After the human EPO gene was cloned in the early 1980s the recombinant form of EPO was rapidly developed and used excessively in clinical practice. First it was used in the management of anemia in patients with chronic renal failure.

Neuroprotective actions of recombinant human erythropoietin (rHu-EPO) have been evaluated both in vitro and in vivo demonstrating antiapoptotic, antioxidative, antiimflammatory, neurotrophic and angiogenic properties [24].

Beneficial effects of EPO in patients with SAH in a clinical trial could not be excluded or concluded on the basis of the study and larger scale trials were warranted [25].

Another phase II randomized double blind placebo controlled trial on intravenous EPO showed that EPO seemed to reduce delayed cerebral ischemia following aSAH via decreasing the severity vasospasm and shortening impaired autoregulation [26].

Lately a phase 1 randomized, parallel, open label study was carried out in healthy volunteers to evaluate the adverse effects of intranasal NeuroEPO. Center for Drug Research and Development (CIDEM, in Spanish) developed a nasal formulation containing EPO with non-hematopoietic activity produced by the Center of Molecular Immunology (CIM, in Spanish). This formulation named NeuroEPO incorporates bioadhesive polymers and other ingredients which increase the residence time in the nasal cavity to enhance its therapeutic effect. This product did not stimulate erytropoesis when it was administered through nasal root [27].

Twenty five volunteers received either the highest dose or half of it. Most of the events were mild, did not require treatment and were completely resolved. No severe adverse events were reported. As a conclusion of this mini trial the authors suggested that the neuroprotective candidate NeuroEpo is a safe product, well tolerated in the nasal mucosa and did not stimulate erythropoiesis in healthy volunteers [28].

5.4.3 Insulin and Insulin Liked Growth Factor-1 (IGF-1)

Insulin and Insulin like Growth Factor (IGF-1) have been shown to be active AKT activators and to be neuroprotective in several stroke models. In preclinical studies insulin has shown promising effects in early brain injury after subarachnoid hemorrhage. Given the fact that intranasal insulin treatment is under study in other CNS disorders, there is no doubt that it can be tested also as a neuroprotective agent in subarachnoid hemorrhage in the future [29, 30].

In conclusion, to date there has been no clinical translation in intranasal treatment after subarachnoid hemorrhage. But given the fact that aSAH is a subgroup of stroke in which cerebral ischemia occurs in a delayed time frame it is likely in the future that promising results in ischemic stroke trials with intranasal treatment of various drugs will be duplicatable in subarachnoid hemorrhage research. Also the aforementioned preclinical studies in aSAH may serve as proof-of-concept data prior to testing the agents in clinical trials.

References

1. Topkoru B, Egemen E, Solaroglu I, Zhang JH. Early brain injury or vasospasm? An overview of common mechanisms. Curr Drug Targets. 2017;18(12):1424–9.
2. Caner B, Hou J, Altay O, Fuj M 2nd, Zhang JH. Transition of research focus from vasospasm to early brain injury after subarachnoid hemorrhage. J Neurochem. 2012;123(Suppl 2):12–21.
3. Connolly ES Jr, Rabinstein AA, Carhuapoma JR, Derdeyn CP, Dion J, Higashida RT, Hoh BL, Kirkness CJ, Naidech AM, Ogilvy CS, Patel AB, Thompson BG, Vespa P, American Heart Association Stroke Council, Council on Cardiovascular Radiology and Intervention, Council on Cardiovascular Nursing; Council on Cardiovascular Surgery and Anesthesia, Council on Clinical Cardiology. Guidelines for the management of aneurysmal subarachnoid hemorrhage: a guideline for healthcare professionals from the American Heart Association/American Stroke Association. Stroke. 2012;43(6):1711–37.
4. Grasso G, Tomasello F. Erythropoietin for subarachnoid hemorrhage: is there a reason for hope. World Neurosurg. 2012;77(1):46–8.
5. Chapman CD, Frey WH 2nd, Craft S, Danielyan L, Hallschmid M, Schiöth HB, Benedict C. Intranasal treatment of central nervous system dysfunction in humans. Pharm Res. 2013;30(10):2475–84.

6. Patel MM, Patel BM. Crossing the blood-brain barrier: recent advances in drug delivery to the brain. CNS Drugs. 2017;31:109–33.
7. Djupesland PG, Messina JC, Mahmoud RA. The nasal approach to delivering treatment for brain diseases: an anatomic, physiologic, and delivery technology overview. Ther Deliv. 2014;5(6):709–33.
8. Thorne RG, Emory CR, Ala TA, Frey WH 2nd. Quantitative analysis of the olfactory pathway for drug delivery to the brain. Brain Res. 1995;692:278–82.
9. Dhuria SV, Hanson LR, Frey WH 2nd. Intranasal delivery to the central nervous system: mechanisms and experimental considerations. J Pharm Sci. 2010;99:1654–73.
10. Suzuki H, Hasegawa Y, Chen W, Kanamaru K, Zhang JH. Recombinant osteopontin in cerebral vasospasm after subarachnoid hemorrhage. Ann Neurol. 2010;68:650–60.
11. Topkoru BC, Altay O, Duris K, Krafft PR, Yan J, Zhang JH. Nasal administration of recombinant osteopontin attenuates early brain injury after subarachnoid hemorrhage. Stroke. 2013;44(11):3189–94.
12. Doyle KP, Yang T, Lessov NS, Ciesielski TM, Stevens SL, Simon RP, et al. Nasal administration of osteopontin peptide mimetics confers neuroprotection in stroke. J Cereb Blood Flow Metab. 2008;28:1235–48.
13. Zhang JH. Vascular neural network in subarachnoid hemorrhage. Transl Stroke Res. 2014;5(4):423–8.
14. Wu J, Zhang Y, Yang P, Enkhjargal B, Manaenko A, Tang J, Pearce WJ, Hartman R, Obenaus A, Chen G, Zhang JH. Recombinant osteopontin stabilizes smooth muscle cell phenotype via integrin receptor/integrin-linked kinase/Rac-1 pathway after subarachnoid hemorrhage in rats. Stroke. 2016;47(5):1319–27.
15. Lau WL, Leaf EM, Hu MC, et al. Vitamin d receptor agonists increase klotho and osteopontin while decreasing aortic calcification in mice with chronic kidney disease fed a high phosphate diet. Kidney Int. 2012;82:1261–70.
16. Enkhjargal B, McBride DW, Manaenko A, Reis C, Sakai Y, Tang J, Zhang JH. Intranasal administration of vitamin D attenuates blood-brain barrier disruption through endogenous upregulation of osteopontin and activation of CD44/P-gp glycosylation signaling after subarachnoid hemorrhage in rats. J Cereb Blood Flow Metab. 2017;37(7):2555–66.
17. Xie Z, Huang L, Enkhjargal B, Reis C, Wan W, Tang J, Cheng Y, Zhang JH. Intranasal administration of recombinant Netrin-1 attenuates neuronal apoptosis by activating DCC/APPL-1/AKT signaling pathway after subarachnoid hemorrhage in rats. Neuropharmacology. 2017;119:123–33.
18. Donega V, Nijboer CH, Braccioli L, Slaper-Cortenbach I, Kavelaars A, van Bel F, Heijnen CJ. Intranasal administration of human MSC for ischemic brain injury in the mouse: in vitro and in vivo neuroregenerative functions. PLoS One. 2014;9(11):e112339.
19. Onda T, Honmou O, Harada K, Houkin K, Hamada H, Kocsis JD. Therapeutic benefits by human mesenchymal stem cells (hMSCs) and Ang-1 gene-modified hMSCs after cerebral ischemia. J Cereb Blood Flow Metab. 2008;28(2):329–40.
20. Khalili MA, Anvari M, Hekmati-Moghadam SH, Sadeghian-Nodoushan F, Fesahat F, Miresmaeili SM. Therapeutic benefit of intravenous transplantation of mesenchymal stem cells after experimental subarachnoid hemorrhage in rats. J Stroke Cerebrovasc Dis. 2012;21:445–51.
21. Khalili MA, Sadeghian-Nodoushan F, Fesahat F, Mir-Esmaeili SM, Anvari M, Hekmati-Moghadam SH. Mesenchymal stem cells improved the ultrastructural morphology of cerebral tissues after subarachnoid hemorrhage in rats. Exp Neurobiol. 2014;23:77–85.
22. Nijboer CH, Kooijman E, van Velthoven CT, van Tilborg E, Tiebosch IA, Eijkelkamp N, Dijkhuizen RM, Kesecioglu J, Heijnen CJ. Intranasal stem cell treatment as a novel therapy for subarachnoid hemorrhage. Stem Cells Dev. 2018;27(5):313–25.
23. Rashed HM, Shamma RN, Basalious EB. Contribution of both olfactory and systemic pathways for brain targeting of nimodipine-loaded lipo-pluronics micelles: in vitro characterization and in vivo biodistribution study after intranasal and intravenous delivery. Drug Deliv. 2016;24(1):181–7.

24. Garcia-Rodiriguez JC, Sosa-Teste I. The nasal route as a potential pathway for delivery of erythropoietin in the treatment of acute ischemic stroke in humans. Sci World J. 2009;9:970–81.
25. Springborg JB, Moller C, Gideon P, Jorgensen OS, Juhler M, Olsen NV. Erythropoietin in patients with aneurysmal subarachnoid haemorrhage: a double blind randomised clinical trial. Acta Neurochir. 2007;149:1089–101.
26. Tseng MY, Hutchinson PJ, Richards HK, Czosnyka M, Pickard JD, Erber WN, Brown S, Kirkpatrick PJ. Acute systemic erythropoietin therapy to reduce delayed ischemic deficits following aneurysmal subarachnoid hemorrhage: a phase II randomized, double-blind, placebo-controlled trial. Clinical article. J Neurosurg. 2009;111:171–80.
27. Muñoz-Cernada A, Pardo-Ruiz Z, Montero-Alejo V, Fernández-Cervera M, Sosa-Testé I, García-Rodríguez JC. Effect of nonionic surfactants and HPMC F4M on the development of formulations of Neuro-EPO as a neuroprotective agent. JAPST. 2014;1:22–35.
28. Santos-Morales O, Díaz-Machado A, Jiménez-Rodríguez D, Pomares-Iturralde Y, Festary-Casanovas T, González-Delgado CA, Pérez-Rodríguez S, Alfonso-Muñoz E, Viada-González C, Piedra-Sierra P, García-García I, Amaro-González D, NeuroEPO Study Group. Nasal administration of the neuroprotective candidate NeuroEPO to healthy volunteers: a randomized, parallel, open-label safety study. BMC Neurol. 2017;17(1):129. https://doi.org/10.1186/s12883-017-0908-0.
29. Grinberg YY, Zitzow LA, Kraig RP. Intranasally administered IGF-1 inhibits spreading depression in vivo. Brain Res. 2017;15(1677):47–57.
30. Zhuang Z, Zhao X, Wu Y, Huang R, Zhu L, Zhang Y, Shi J. The anti-apoptotic effect of PI3K-Akt signaling pathway after subarachnoid hemorrhage in rats. Ann Clin Lab Sci. 2011;41(4):364–72.

Chapter 6
Intranasal Delivery of Therapeutic Peptides for Treatment of Ischemic Brain Injury

Tingting Huang, Amanda Smith, Jun Chen, and Peiying Li

Abstract There is an unmet need in the treatment of cerebral ischemic stroke to enhance post-stroke functional recovery. Intranasal delivery of therapeutic peptides has been emerging as an important strategy to improve stroke recovery. In this chapter, we introduce the definition and mechanisms of intranasal delivery of therapeutic peptides. We also discuss its advantages and disadvantages in the treatment of stroke. A variety of peptides and the administration regimens that have been tested in stroke animal models are listed. We believe that further investigation in this regard can deepen our understanding of intranasal delivery and may promote its clinical translation in the pursuit of better stroke recovery.

Keywords Intranasal delivery · Therapeutic peptides · Ischemic stroke · Brain recovery

6.1 Introduction

Stroke is the second leading cause of death and third main reason for severe disabilities worldwide [1, 2]. Over 80% of stroke cases present as ischemic stroke [3]. Targeting the highly dynamic events that occur during ischemic stroke in the relatively inaccessible brain microenvironment remains a challenge.

After decades of research, a variety of peptides, including erythropoietin (EPO), interleukin 4 (IL-4), and transforming growth factor (TGF)-β1, have emerged as effective therapeutic agents to treat brain diseases, such as neurodegeneration, pain, psychiatric disorders and stroke [4]. However, some peptides cannot pass the blood brain barrier due to their high molecular weight, thereby requiring some form of invasive delivery to access the brain [5]. Recently, intranasal delivery has gained attention as a novel non-invasive drug delivery route. In this chapter, we will define intranasal delivery, describe its possible mechanisms, and discuss its advantages and disadvantages as a delivery method of therapeutic peptides. Some peptides that have shown protection against stroke through intranasal delivery along with the regimens that have been used, will also be covered in this chapter.

6.2 Definition and Mechanisms of Intranasal Delivery of Therapeutic Peptides

Intranasal delivery is an alternative drug delivery strategy used to treat brain injuries, such as stroke [6]. Intranasal administration circumvents the blood brain barrier (BBB) and provides a direct and rapid route for drugs to enter the brain and cerebrospinal fluid (CSF) through the olfactory epithelium [7]. Because of the large absorptive surface area of the nasal cavity (~ 160 cm^2) [8], along with its high vascularization and porous epithelium, drugs or treatment agents delivered through the intranasal route can enter the brain within minutes [9]. Chemicals, peptides, genes, and cells have been successfully delivered to the brain by intranasal delivery to achieve neuroprotection [10].

After intranasal administration of [^{125}I]-labeled proteins to rats and monkeys, radiolabeling has been shown to extend from the olfactory and trigeminal nerve components in the nasal epithelium to the olfactory bulb and brainstem, respectively. The signals from these [^{125}I]-labeled proteins can spread to other central nervous system (CNS) areas from these initial sites [11, 12]. The olfactory region is adjacent to the CSF flow tracts around the olfactory lobe [13]. Therefore, intranasal administration could lead to direct delivery into the CSF.

Two possible mechanistic routes have been suggested to underlie direct brain delivery after intranasal administration: (1) extracellular routes and (2) intracellular routes [7]. The extracellular route is a rapid pathway by which the drug is absorbed across olfactory epithelial cells, either by transcellular or paracellular mechanisms. Then the agent can be taken up into the CNS directly [7], which explains why the delivery of intranasal drugs to the brain occurs within minutes [14]. On the other hand, during the intracellular route, drugs are internalized into primary neurons of the olfactory epithelium by endocytosis, which includes pinocytotic mechanisms, followed by intracellular axonal transport to the olfactory bulb [7]. This intracellular pathway may take hours for drugs to be dispersed within the brain [15]. The delivery route and mechanisms of intranasal drug absorption is illustrated in Fig. 6.1.

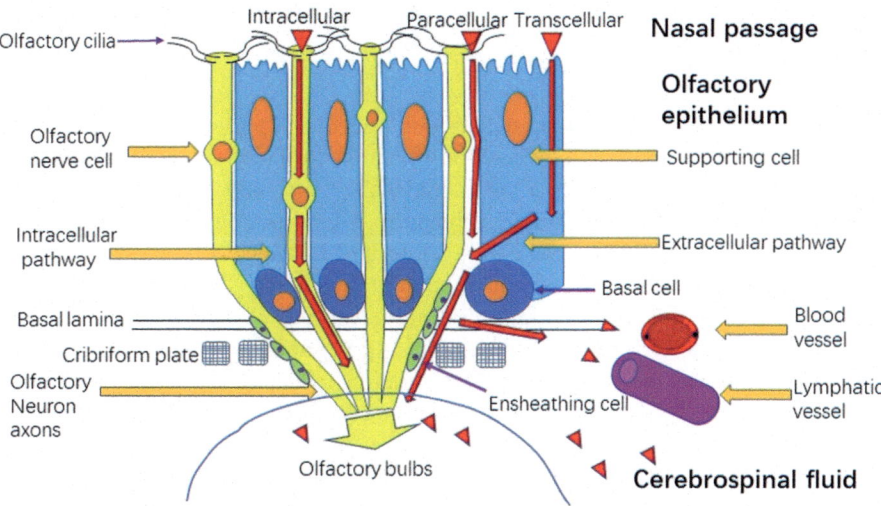

Fig. 6.1 Intranasal delivery pathways across the olfactory epithelium are depicted. Potential transport routes are shown in red. Some agents may be delivered via an intracellular pathway from the olfactory epithelium to the olfactory bulb by olfactory sensory neurons by endocytic, including pinocytotic, mechanisms, followed by intracellular axonal transport to the olfactory bulb. Others agents may pass the olfactory epithelial barrier by paracellular or transcellular transport to reach the lamina propria, and then distribute to the brain in multiple ways: (1) via entry into olfactory blood vessels and the systemic circulation; (2) by absorption into olfactory lymphatic vessels, which leads to the deep cervical lymph nodes of the neck; (3) by diffusion to the subarachnoid space with CSF, or. (4) entry into channels created by olfactory enshreathing cells surrounding the olfactory nerves, where they can access the olfactory bulbs. Subsequently, intranasally delivered drugs are transported from the olfactory bulb to the CSF surrounding the whole brain

6.3 Advantages of Intranasal Delivery of Therapeutic Peptides

Compared to conventional delivery methods, intranasal delivery of therapeutic peptides possesses the following advantages; (1) Safe and less invasive [16] compared to local delivery into the brain, which usually requires intraventricular and parenchymal injections [17, 18]; (2) easier passage into the brain [16]; (3) less requirement for the high diffusibility of therapeutic agents [16]; (4) lower dosage can be used with more reliable pharmacological profiles due to less drug metabolism and degradation along the delivery routes [19, 20]; (5) faster onset of action [21]; (6) less systemic side effects due to the absence of intestinal absorption and degradation by the enterohepatic cycle as well as general distribution [9, 22]; (7) large molecules can gain entry into the brain which is not easily attainable via the oral route [23]; and (8) less requirement for sterile preparation [24].

6.4 Peptides that Have Been Administered Intranasally to Treat Stroke

Recently, there has been increased focus on specific peptides as neuroprotective agents in experimental models of cerebral ischemic stroke. For example, numerous studies have shown that EPO administered subcutaneously, intraperitoneally, or through intra-cerebroventricular (ICV) injection, is effective against ischemic stroke [25–27]. Subsequently, intranasal administration of EPO, as well as other peptides, has shown efficacy in experimental models of ischemic stroke. The protective effects of these peptides mainly fall into four categories: (1) inhibition of neuronal apoptosis, as observed with EPO treatment [28]; (2) amelioration of inflammatory responses, as observed with Tat-NEMO-binding domain (Tat-NBD), a stable-mutant form of plasminogen activator inhibitor-I (CAPI) [20, 29]; (3) promotion of progenitor cell proliferation and brain repair mechanisms, as detected with insulin-like growth factor-I (IGF-1), TGF-β1, and TGF-α; [30–33] and (4) improvement in white matter function by preventing demyelination and promoting axon sprouting, as observed with tissue plasminogen activator (tPA) and Apo-transferrin (aTf) [31, 34]. Table 6.1 summarizes the neuroprotective effects of intranasal administration of these peptides against cerebral ischemic stroke.

In addition to the above-mentioned peptides, there are numerous peptides that have shown to be effective regulators of ischemic brain injury through ICV or subcutaneous administration, such as IL-4 and IL-33 [41, 42]. The efficacy of intranasal delivery of these peptides against ischemic stroke has not been examined. However, they are promising candidates for this new route of administration. Thus further studies investigating the efficacy of intranasal delivery of these two peptides against ischemic stroke is warranted.

6.5 Administration Regimens of Intranasal Delivery in the Treatment of Stroke

One of the advantages of intranasal delivery is that it reduces the effective dosage compared to systemic administration. However, there are still tremendous differences regarding the dosages reported by different studies for different peptides. In addition, some studies also report multiple intranasal peptide administrations or combination of different peptides. In Table 6.2, we summarize the intranasal infusion protocols used to administer various peptides and their corresponding experimental ischemic animal model.

Table 6.1 Neuroprotective effect of intranasal delivery of various peptides against stroke

Peptide	Therapeutic effects and possible mechanisms	References
EPO	Increased cell viability and decreased the number of non-viable cells	[28]
Tat-NBD	Attenuated NF-κB signaling, microglia activation and brain damage	[20]
IGF-1	Reduced infarct volume, hemispheric swelling, and promoted proliferation of neuronal progenitor cells	[30, 31]
tPA	Rescued long-term spatial memory, prevented demyelination, and loss of axonal conduction	[34]
TGF-β1	Decreased infarct volume, improved functional recovery, and enhanced neurogenesis	[32]
CAPI	Reduced acute LPS/HI-triggered NF-κB activity, pro-inflammatory IL-6 production, and brain tissue loss	[29]
TGF-α	Triggered the proliferation of neuronal progenitor cells and significantly improved behavioral response	[33]
aTf	Reduced white matter damage and accelerated remyelination	[31]
Apelin-13	Reduced microglia activation, attenuated inflammatory cytokines/chemokines levels, and decreased apoptotic cell death. Suppressed caspase-3 activation and increased the survival gene Bcl-2 after stroke	[35]
CART	Upregulated BDNF expression. Enhanced survival, proliferation and migration of NPCs in SVZ. Improved neurological function and facilitated neural regeneration	[36]
HBHP	Reduced NCM-mediated aggravation of neuronal death induced by sublethal concentrations of NMDA or zinc. Improved motor impairment and neurological deficits	[37]
bFGF	Improved the three key clinical indicators of cerebral I/R injury severity: The neurologic deficit score, locomotor activity and infarct volume	[38]
Exendin-4	Reduced neurological deficit scores and infarct volume	[39]
G-CSF	Decreased infarct volume and improved neurological functional recovery. Upregulated HO-1 and reduced calcium overload following ischemia. Increased cerebrovascular density and newborn cell numbers after brain ischemia	[40]

EPO erythropoietin, *Tat-NBD* Tat-NEMO-binding domain, cell-penetrating anti-NF-κB peptides, *IGF-1* insulin-like growth factor-I, *CAPI* a stable-mutant form of Plasminogen activator inhibitor-I, *tPA* tissue plasminogen activator, *TGF* transforming growth factor, *CART* cocaine-and amphetamine-regulated transcript peptide, *HBHP* HMGB1 binding heptamer peptide, *bFGF* basic fibroblast growth factor, *G-CSF* granulocyte colony-stimulating factor, *HI* hypoxic-ischemic, *LPS* lipopolysaccharide, *IL-6* interleukin-6, *BDNF* brain derived neurotrophic factor, *SVZ* subventricular zone, *NPC* neural progenitor cells, *HO-1* heme oxygenase-1, *I/R* ischemia-reperfusion, *NMDA* N-methyl-D-aspartate, *NCM* NMDA-conditioned medium, *NF-κB* nuclear factor-κB

6.6 Disadvantages and Limitations of Intranasal Delivery of Therapeutic Drugs

Despite the well-known advantages of intranasal drug delivery as described above, intranasal delivery of therapeutic drugs also carries some disadvantages: (1) Potentially rapid degradation of the drug due to active mucociliary clearance and the presence of nasal cytochrome P450/peptidases/proteases. This disadvantage can

Table 6.2 Administration regimens of intranasal delivery of therapeutic peptides

Peptide	Dosage	Single / Multiple (Times)	Combination peptide	Testing model	Effects	Reference
rhEPO	1.2 U	S	/	Rat MCAO	Not significant	[43]
	4.8 U	S	/		Most effective	
	12 U	S	/		Most effective	
	24 U	S	/		Weaker effects	
Neuro-EPO	249.4 UI/10 µL	M	/	Gerbil MCAO	Effective	[44]
Tat-NBD	1.4 mg/kg	S	/	Pup HI	Marked effects	[20]
	5.6 mg/kg	S	/		Little protection	
tPA	300 µg	M (4)	/	Mice MCAO	Significant	[45]
	600 µg	S	/	Rat MCAO	Significant	[46]
	0.065 µg/mL	S	/	Cultured cortical neurons	Significant	[45]
	0.65 µg/mL	S	/		Significant	
	2 µg/mL	S	/		Significant	
	6.5 µg/mL	S	/		Peak	
	13 µg/mL	S	/		Reduce	
IGF-1	12 µL	S	EPO	Mice MCAO	Significant	[47]

rhEPO recombinant human erythropoietin, *Tat-NBD* cell-penetrating anti-NF-κB peptides, *tPA* tissue plasminogen activator, *IGF-1* insulin-like growth factor-I, *MCAO* middle cerebral artery occlusion, *HI* hypoxic-ischemic

be overcome by covalently binding the drug to polyethylene glycol (PEG), also known as PEGylation [48]. PEGylation has been used to enhance drug delivery of BDNF into the brain, where it reduced systemic BDNF clearance and facilitated neuroprotection [49, 50]. Intranasal delivery of PEGylated-TGF-α induced similar effects to intracranial delivery of TGF-α, such as inducing neural stem cells in the ependymal layer and progeny in the subventricular zone to proliferate, migrate to the lesion area, and differentiate into context-appropriate neurons [51]. However, intranasal infusion of un-PEGylated TGF-α failed to produce similar effects, which may be attributed to its vulnerability to degradation [33]. (2) Although intranasal delivery can bypass the BBB, the major disadvantage of this route is the limited absorption across the nasal epithelium. This has restricted its application to particularly potent substances. Hyaluronidase, an enzyme that binds to hyaluronic acid, can reduce the viscosity of hyaluronic acid or tissue [52]. Therefore, application of hyaluronidase has become a routine procedure used to facilitate intranasal drug delivery. Using 100 U hyaluronidase 30 min before apelin-13 administration increased its tissue permeability. It helped to reduce microglia activation, decreased the levels of inflammatory cytokines or chemokines and inhibited neuronal apoptosis [35]. However, there is also evidence that without hyaluronidase, apelin-13 alone can produce similar protective effects [53–55]. Therefore, further studies should be carried out to determine whether there is actual benefit to applying hyaluronidase. (3) Because of the size of the nasal cavity, there is a limit to the drug volume that can be delivered, which ranges from 25 to 200 μL. Indeed, in order to avoid discomfort, 2 μL drops are typically infused into one nasal cavity at a time, alternating between nostrils every 2 min, which takes over 20 min to complete drug administration.

6.7 Future Directions and Concluding Remarks

As a non-invasive approach, intranasal delivery of therapeutic peptides has numerous advantages, including easier passage to the brain, reduced dosage, and faster onset of action. Thus it is emerging as a promising alternative route for drug administration in the treatment of cerebral ischemic stroke. Multiple therapeutic peptides, including EPO, tPA, TGF-β1 have been shown to protect against stroke when administered intranasally. However, there are a lot of unknowns pertaining to the delivery details, such as the concentration that is needed, the number of injections required, and the times at which the drug should be delivered. Further investigation into intranasal delivery of therapeutic peptides as a neuroprotective strategy may lead to an easily accessible stroke treatment strategy, and hence better recovery after stroke.

References

1. Chen R, et al. Decreased percentage of peripheral naive T cells is independently associated with ischemic stroke in patients on hemodialysis. Int Urol Nephrol. 2017;49(11):2051–60.
2. Whiteford HA, et al. Global burden of disease attributable to mental and substance use disorders: findings from the Global Burden of Disease Study 2010. Lancet. 2013;382(9904):1575–86.
3. Mozaffarian D, et al. Heart disease and stroke statistics--2015 update: a report from the American Heart Association. Circulation. 2015;131(4):e29–322.
4. Strand FL. Neuropeptides: general characteristics and neuropharmaceutical potential in treating CNS disorders. Prog Drug Res. 2003;61:1–37.
5. Lalatsa A, Schatzlein AG, Uchegbu IF. Strategies to deliver peptide drugs to the brain. Mol Pharm. 2014;11(4):1081–93.
6. Ma M, et al. Intranasal delivery of transforming growth factor-beta1 in mice after stroke reduces infarct volume and increases neurogenesis in the subventricular zone. BMC Neurosci. 2008;9:117.
7. Illum L. Is nose-to-brain transport of drugs in man a reality? J Pharm Pharmacol. 2004;56(1):3–17.
8. Harkema JR, Carey SA, Wagner JG. The nose revisited: a brief review of the comparative structure, function, and toxicologic pathology of the nasal epithelium. Toxicol Pathol. 2006;34(3):252–69.

9. Fortuna A, et al. Intranasal delivery of systemic-acting drugs: small-molecules and biomacromolecules. Eur J Pharm Biopharm. 2014;88(1):8–27.
10. Lochhead JJ, Thorne RG. Intranasal delivery of biologics to the central nervous system. Adv Drug Deliv Rev. 2012;64(7):614–28.
11. Thorne RG, et al. Delivery of insulin-like growth factor-I to the rat brain and spinal cord along olfactory and trigeminal pathways following intranasal administration. Neuroscience. 2004;127(2):481–96.
12. Thorne RG, et al. Delivery of interferon-beta to the monkey nervous system following intranasal administration. Neuroscience. 2008;152(3):785–97.
13. Pardridge WM. Drug transport across the blood-brain barrier. J Cereb Blood Flow Metab. 2012;32(11):1959–72.
14. Thorne RG, Frey WH 2nd. Delivery of neurotrophic factors to the central nervous system: pharmacokinetic considerations. Clin Pharmacokinet. 2001;40(12):907–46.
15. Shipley MT. Transport of molecules from nose to brain: transneuronal anterograde and retrograde labeling in the rat olfactory system by wheat germ agglutinin-horseradish peroxidase applied to the nasal epithelium. Brain Res Bull. 1985;15(2):129–42.
16. Rhim T, Lee DY, Lee M. Drug delivery systems for the treatment of ischemic stroke. Pharm Res. 2013;30(10):2429–44.
17. Wang Y, et al. Hydrogel delivery of erythropoietin to the brain for endogenous stem cell stimulation after stroke injury. Biomaterials. 2012;33(9):2681–92.
18. Chen MY, et al. Surface properties, more than size, limiting convective distribution of virus-sized particles and viruses in the central nervous system. J Neurosurg. 2005;103(2):311–9.
19. Thompson BJ, Ronaldson PT. Drug delivery to the ischemic brain. Adv Pharmacol. 2014;71:165–202.
20. Yang D, et al. Intranasal delivery of cell-penetrating anti-NF-kappaB peptides (tat-NBD) alleviates infection-sensitized hypoxic-ischemic brain injury. Exp Neurol. 2013;247:447–55.
21. Saver JL, et al. Time to treatment with intravenous tissue plasminogen activator and outcome from acute ischemic stroke. JAMA. 2013;309(23):2480–8.
22. Pires A, et al. Intranasal drug delivery: how, why and what for? J Pharm Pharm Sci. 2009;12(3):288–311.
23. Ozsoy Y, Gungor S, Cevher E. Nasal delivery of high molecular weight drugs. Molecules. 2009;14(9):3754–79.
24. Costantino HR, et al. Intranasal delivery: physicochemical and therapeutic aspects. Int J Pharm. 2007;337(1–2):1–24.
25. Iwai M, et al. Erythropoietin promotes neuronal replacement through revascularization and neurogenesis after neonatal hypoxia/ischemia in rats. Stroke. 2007;38(10):2795–803.
26. Iwai M, et al. Enhanced oligodendrogenesis and recovery of neurological function by erythropoietin after neonatal hypoxic/ischemic brain injury. Stroke. 2010;41(5):1032–7.
27. Zhang F, et al. Enhanced delivery of erythropoietin across the blood-brain barrier for neuroprotection against ischemic neuronal injury. Transl Stroke Res. 2010;1(2):113–21.
28. Castaneda-Arellano R, Feria-Velasco AI, Rivera-Cervantes MC. Activity increase in EpoR and Epo expression by intranasal recombinant human erythropoietin (rhEpo) administration in ischemic hippocampi of adult rats. Neurosci Lett. 2014;583:16–20.
29. Yang D, et al. Taming neonatal hypoxic-ischemic brain injury by intranasal delivery of plasminogen activator inhibitor-1. Stroke. 2013;44(9):2623–7.
30. Liu XF, et al. Non-invasive intranasal insulin-like growth factor-I reduces infarct volume and improves neurologic function in rats following middle cerebral artery occlusion. Neurosci Lett. 2001;308(2):91–4.
31. Guardia Clausi M, et al. Inhalation of growth factors and apo-transferrin to protect and repair the hypoxic-ischemic brain. Pharmacol Res. 2016;109:81–5.
32. Mishra L, Derynck R, Mishra B. Transforming growth factor-beta signaling in stem cells and cancer. Science. 2005;310(5745):68–71.
33. Guerra-Crespo M, et al. Intranasal administration of PEGylated transforming growth factor-alpha improves behavioral deficits in a chronic stroke model. J Stroke Cerebrovasc Dis. 2010;19(1):3–9.

34. Xia Y, et al. Tissue plasminogen activator promotes white matter integrity and functional recovery in a murine model of traumatic brain injury. Proc Natl Acad Sci U S A. 2018;115(39):E9230–8.
35. Chen D, et al. Intranasal delivery of apelin-13 is neuroprotective and promotes angiogenesis after ischemic stroke in mice. ASN Neuro. 2015;7(5):1759091415605114. https://doi.org/10.1177/1759091415605114.
36. Luo Y, et al. CART peptide induces neuroregeneration in stroke rats. J Cereb Blood Flow Metab. 2013;33(2):300–10.
37. Kim ID, et al. Intranasal delivery of HMGB1-binding heptamer peptide confers a robust neuroprotection in the postischemic brain. Neurosci Lett. 2012;525(2):179–83.
38. Zhao YZ, et al. Intranasal delivery of bFGF with nanoliposomes enhances in vivo neuroprotection and neural injury recovery in a rodent stroke model. J Control Release. 2016;224:165–75.
39. Zhang H, et al. Intranasal delivery of exendin-4 confers neuroprotective effect against cerebral ischemia in mice. AAPS J. 2016;18(2):385–94.
40. Sun BL, et al. Intranasal delivery of granulocyte colony-stimulating factor enhances its neuroprotective effects against ischemic brain injury in rats. Mol Neurobiol. 2016;53(1):320–30.
41. Zhao X, et al. Neuronal interleukin-4 as a modulator of microglial pathways and ischemic brain damage. J Neurosci. 2015;35(32):11281–91.
42. Yang Y, et al. ST2/IL-33-dependent microglial response limits acute ischemic brain injury. J Neurosci. 2017;37(18):4692–704.
43. Yu YP, et al. Intranasal recombinant human erythropoietin protects rats against focal cerebral ischemia. Neurosci Lett. 2005;387(1):5–10.
44. Teste IS, et al. Dose effect evaluation and therapeutic window of the neuro-EPO nasal application for the treatment of the focal ischemia model in the Mongolian gerbil. ScientificWorldJournal. 2012;2012:607498.
45. Chen N, et al. Subacute intranasal administration of tissue plasminogen activator improves stroke recovery by inducing axonal remodeling in mice. Exp Neurol. 2018;304:82–9.
46. Liu Z, et al. Subacute intranasal administration of tissue plasminogen activator increases functional recovery and axonal remodeling after stroke in rats. Neurobiol Dis. 2012;45(2):804–9.
47. Fletcher L, et al. Intranasal delivery of erythropoietin plus insulin-like growth factor-I for acute neuroprotection in stroke. Laboratory investigation. J Neurosurg. 2009;111(1):164–70.
48. Fishburn CS. The pharmacology of PEGylation: balancing PD with PK to generate novel therapeutics. J Pharm Sci. 2008;97(10):4167–83.
49. Sakane T, Pardridge WM. Carboxyl-directed pegylation of brain-derived neurotrophic factor markedly reduces systemic clearance with minimal loss of biologic activity. Pharm Res. 1997;14(8):1085–91.
50. Wu D, Pardridge WM. Neuroprotection with noninvasive neurotrophin delivery to the brain. Proc Natl Acad Sci U S A. 1999;96(1):254–9.
51. Guerra-Crespo M, et al. Transforming growth factor-alpha induces neurogenesis and behavioral improvement in a chronic stroke model. Neuroscience. 2009;160(2):470–83.
52. Clement WA, et al. The use of hyaluronidase in nasal infiltration: prospective randomized controlled pilot study. J Laryngol Otol. 2003;117(8):614–8.
53. O'Donnell LA, et al. Apelin, an endogenous neuronal peptide, protects hippocampal neurons against excitotoxic injury. J Neurochem. 2007;102(6):1905–17.
54. Zeng XJ, et al. Neuroprotective effect of the endogenous neural peptide apelin in cultured mouse cortical neurons. Exp Cell Res. 2010;316(11):1773–83.
55. Yang Y, et al. Apelin-13 protects the brain against ischemia/reperfusion injury through activating PI3K/Akt and ERK1/2 signaling pathways. Neurosci Lett. 2014;568:44–9.

Chapter 7
Intranasal Delivering Method in the Treatment of Ischemic Stroke

Chunhua Chen, Mengqin Zhang, Yejun Wu, Changman Zhou, and Renyu Liu

Abstract Ischemic stroke is a leading cause of death and disability worldwide. Advances in early recognition of stroke symptoms and the transport of patients to specialized stroke centers has been a major step in improvement of mortality and morbidity. Speed is essential since current intravenous thrombolytic treatments can only be delivered in a very narrow therapeutic window. Intravenous therapy requires specialized skills, subjects the medication to first pass metabolism, and the issue of blood to brain transport is a major problem.

An alternate approach, the intranasal route, could deliver medication to the target in a quick manner and overcome the blood brain barrier to the central nervous system while avoiding first pass metabolism. Intranasal medication also requires minimal skill to administer in a hospital or in the field.

This chapter will address the pathway through which substances travel from the nasal epithelium to various regions of the central nervous system. This includes multiple substances for intranasal administration for the potential treatment of ischemic stroke, such as proteins and peptides, stem cells, gene vectors and nanoparticles. The chapter will conclude with the merits and potential issues of intranasal administration, as well as future directions.

C. Chen · C. Zhou
Department of Anatomy and Embryology, Peking University Health Science Center, Beijing, China

M. Zhang · Y. Wu
School of Basic Medical Sciences, Peking University Health Science Center, Beijing, China

R. Liu (✉)
Department of Anesthesiology and Critical Care, Perelman School of Medicine, University of Pennsylvania, Philadelphia, PA, USA
e-mail: RenYu.Liu@pennmedicine.upenn.edu

Keywords Blood brain barrier · Central nervous system · Intranasal administration · Stroke

Ischemic stroke is still one of the leading causes of death across the world. Thrombolytic therapy is available for ischemic stroke management, which can only be delivered in a very narrow therapeutic window within 4.5 h [1]. Neuroprotective agents have been tested effective in various animal models, however, none of them was able to achieve to be a clinical neuroprotectant. One of the potential causes is that these agents can only be delivered to the stroke patients after the arrival to the hospital, losing the opportunity to be delivered in a quicker manner. To develop an intranasal approach could potentially overcome some of the barriers to deliver the medication to the target in a quick manner especially when there is no intravenous route available. Another major benefit to develop the intranasal formulation is that it could potentially overcome the blood brain barrier (BBB) to deliver the macromolecules to the central nervous system (CNS) when the traditional intravenous approach will not work at all. It may also bypass the first pass effect from intranasal administration. To avoid intravenous administration disadvantages, intranasal administration has other critical benefits including lack or reduction of systemic side effects and increasing the efficacy by avoiding metabolism or hydrolysis in the blood for certain medications.

7.1 Pathways for Intranasal Administration

The pathway through which substances travel from the nasal epithelium to various regions of the CNS has not been fully investigated. However, previous studies reported that intranasally delivered macromolecules could bypass the BBB to elicit biological and pharmacological effects rapidly in the brain and spinal cord in rats and monkeys [2, 3]. There are at least three sequential transport steps as described as follows.

7.1.1 Transport Across the Olfactory or Respiratory Epithelial Barriers

Substances intranasally administered initially crossed the olfactory or respiratory epithelium either by intracellular or extracellular pathways. Intracellular pathways across the olfactory or respiratory epithelium include endocytosis into olfactory sensory neuron (OSN) or trigeminal neuron processes and subsequent intracellular transport to the olfactory bulb or the brainstem [4–12]. As for the extracellular pathways, substances cross the epithelium to assess the underlying lamina propria potentially by paracellular diffusion. Horseradish peroxidase (HRP) has been shown to reach the olfactory bulb not only through intracellular uptake by OSN, but also by passing through open intercellular clefts [13]. The expression of the tight junction (TJ) proteins in the epithelium significantly determines the paracellular permeability [14, 15].

7.1.2 Transport from the Nasal Lamina Propria to Sites of Brain Entry

After entering the nasal laminal propria across the olfactory and respiratory epithelium via intracellular and extracellular pathways mentioned above, substances are to reach the brain entry sites (olfactory bulbs and brainstem) through both olfactory and trigeminal nerves. This transport may via intracellular pathways and extracellular pathways. The former is endocytosis and intraneuronal transport with OSN or trigeminal ganglion cells, while the latter happens when substances diffuse within perineural, perivascular or lymphatic channels associated with olfactory and trigeminal nerve bundles extending from the lamina propria to the olfactory bulbs and brainstem, respectively [5–7]. Fates of substances reaching the extracellular environment in the lamina propria are different and includes: (a) absorption into blood vessels and entry to systemic circulation; (b) absorption into lymphatic vessels and entry to deep cervical lymphatic nodes; (c) diffusion around nerve bundles and entry to brain. However, the travel time of different mechanisms varies. Some researchers have simulated and estimated the travel time of these mechanisms. It seems that the extracellular bulk convection is the most plausible mechanism for the rapid transport [16]. It takes at most 30 min, which fits the experimental results well.

7.1.3 Transport from the Brain Entry Sites to Widespread Sites Within the CNS

Substances may be distributed to widespread sites of CNS from the entry sites through the bulk flow in the perivascular spaces of cerebral blood vessels [17, 18]. Many studies attempted to elucidate the flow's direction and characteristics. However, different groups get different results [17–19]. It has been shown that rats with high blood pressure and heart rate displayed a larger distribution of adeno-associated virus 2, fluorescent liposomes, and bovine serum albumin, which suggests that fluid circulation within the CNS through the perivascular space is the primary mechanism [20]. What's more, following intranasal administration, uptake of [^{125}I]-calcitonin and [^{125}I]-erythropoietin (EPO) are abolished with surgical transection of the rostral migratory stream (RMS), the pathway along which neuronal precursors migrate to the olfactory bulb. This provides evidence of the vital role of the RMS in the CNS delivery of intranasally administered agents [21]. In addition, some of the substances could be lost in the nasal-associated lymphoid tissues [22]. Another study indicates that oxytocin (OT) reached the CSF within 1 h after intranasal administration [23]. Following tracer application, substances rapidly enter into the CSF along perivascular spaces [24, 25] while limited distribution along perivascular spaces [26, 27]. The mechanism about the perivascular distribution following intranasal administration to target requires further research.

7.2 Multiple Substances for Intranasal Administration for Ischemic Stroke

Studies have demonstrated that intranasal delivery of multiple substances can effectively prevent the ischemic brain injury in animal models. The substances can be divided into four types, which were listed in Table 7.1: (1) proteins and peptides; (2) stem cells; (3) gene vectors; (4) nanoparticles.

7.2.1 Proteins and Peptides

Numerous proteins and peptides have been suggested to have a robust neuroprotective effect in focal cerebral ischemia via intranasal administration.

Administration of insulin-like factors (IGF-1) by the intranasal approach has shown significantly reduced infarct volume and improved motor-sensory and somatosensory functions in rats. Also, intranasal IGF-1 after middle cerebral artery occlusion (MCAO) decreased neuronal apoptosis in the ischemic ipsilateral hemisphere [28]. Intranasal delivery of granulocyte colony-stimulating factor (G-CSF) also decreased infarct volume, increased recovery of neurological function and promoted angiogenesis and neurogenesis following ischemia in rats [31]. What's more, study has provided evidence that intranasal administration of exendin-4 exerted a neuroprotective effect mediated by anti-apoptotic mechanism in MCAO mice and protected neurons against ischemic injury through the glucagon-like peptide 1 receptor (GLP-1R) pathway [32]. In addition, high mobility group box 1 (HMGB1) binding heptamer peptide [33], tissue plasminogen activator (tPA) [34, 35], exogenous recombinant human erythropoietin (rHu-Epo) [36], Neuro-erythropoietin (Neuro-EPO) [38, 39], transforming growth factor-beta1 [40], Wnt-3a [41], Apelin-13 [42], nerve growth factor [43] also have a positive impact on ischemic stroke through the intranasal administration.

7 Intranasal Delivering Method in the Treatment of Ischemic Stroke

Table 7.1 The intranasal delivery of multiple substances on animal models of ischemic stroke

Catalog	Substances	Functions	Animal model	References
Peptide and protein	IGF-1 and insulin	Reduced infarct volume Decreased apoptosis Improved neurologic functions	Rats and mice MCAO	[28–30]
	G-CSF	Reduced neuronal damage Promoted angiogenesis and neurogenesis	Rats MCAO	[31]
	Exendin-4	Exerted neuroprotective efforts by anti-apoptotic mechanism	Mice MCAO	[32]
	HBHP	Inhibit HMGB1 to reduce damages	Rats MCAO	[33]
	Recombinant tPA	Improved nervous functions Reduced the cortical stimulation threshold Enhanced neurogenesis increased the level of mature brain-derived neurotrophic factors	Rats TBI and MCAO	[34, 35]
	rHu-Epo	Promoted neuroprotection	Rats chronic hypoxia	[36, 37]
	Neuro-EPO	Reduced delayed neuronal death	Mongolian gerbil CCAO	[38, 39]
	TGF-beta1	Reduced infarct volume Improved functional recovery Enhanced neurogenesis	Mice MCAO	[40]
	Wnt-3a	Reduced infarct volume by enhancing the cerebral blood flow Enhanced neurogenesis	Mice MCAO	[41]
	Apelin-13	Reduced inflammation Decreased cell death Increased angiogenesis	Mice MCAO	[42]
	NGF	Reduced infarct volume Improved neurologic function	Rats MCAO	[43]
	PAI-1	Reduced the extravascular toxicity of tPA Reduced brain atrophy Attenuated neuroinflammation	Rat pups HI	[44]
	Caspase-9 inhibitor	Reduced neuron death	Rats MCAO	[45]
	IL-1RA	Promoted neuroprotection	Rats MCAO	[46]
	OPN	Reduced infarct volume Reduced inflammation Ameliorated neurological deficits	Rats MCAO	[47, 48]

(continued)

Table 7.1 (continued)

Catalog	Substances	Functions	Animal model	References
Stem cells	BMSC	Reduced infarct volume Induced long-lasting cell proliferation	Rats; Mice [49] MCAO	[49–51]
Gene vector	HMGB1 siRNA	Inhibited HMGB1 to reduce damages	Rats MCAO	[52]
	iNOS siRNA	Reduced infarct volume Reduced neurologic deficits	Rats MCAO	[53]
Small molecule and others	Salvinorin A	Protected the brain and improved neurological outcome	Mice MCAO	[54]
	Polycation-shielded $Ca^{(2+)}$/nucleotide nanocomplexes	Reduced infarct volume	Rat MCAO	[55]
	bFGF nanoliposomal	Improved accumulation of bFGF	Rats MCAO	[56]
	Progesterone	Decreased the mortality rate Improved motor function Reduced infarct volume Decreased the early BBB disruptions	Mice MCAO	[57]
	GRb1	Reduced infarct volume	Rats MCAO	[58]
	DFO	Reduced infarct volume	Rats MCAO	[59]
	CPA	Decreased ischemic damages	Rats MCAO	[60]
	Z-LIG	Enhanced protection against ischemic injury	Rats MCAO	[61]
	Xingnaojing mPEG2000-PLA modified microemulsion	Used for ischemic treatment	Mice MCAO	[62]

IGF-1 insulin-like growth factor 1, *G-CSF* granulocyte colony-stimulating factor, *HBHP* HMGB1 binding heptamer peptide, *HMGB1* high mobility group box 1, *tPA* tissue plasminogen activator, *rHu-EPO* recombinant human erythropoietin, *TGF-beta1* transforming growth factor-beta1, *NGF* nerve growth factor, *PAI-1* plasminogen activator inhibitor-I, *IL-1RA* interleukin-1 receptor antagonist, *BMSCs* bone marrow mesenchymal stem cells, *OPN* osteopontin, *bFGF* basic fibroblast growth factors, *GRb1* Ginsenoside Rb1, *DFO* deferoxamine, *CPA* N(6)-cyclopentyladenosine, *Z-LIG* Z-Ligustilide, *XNJ-M* xingnaojing microemulsion, *TBI* traumatic brain injury, *CCAO* common carotid artery occlusion, *HI* hypoxia and ischemia

7.2.2 Stem Cells

Stem cell intranasal delivery is a promising treatment possibility for ischemic stroke due to their potential ability to deliver neurotrophic factors to damaged cells. Intranasal bone marrow mesenchymal stem cells (BMSCs) transplantation after neonatal stroke in rats has neuroprotection and great potential as a regenerative therapy to enhance neurovascular regeneration and improve functional recovery observed at the juvenile stage of development [50]. Meanwhile, mesenchymal stem cells (MSCs) and MSC over-expressing brain-derived neurotrophic factor (MSC-BDNF) significantly reduced infarct size and gray matter loss and induced long-lasting cell proliferation in the ischemic hemisphere in rats [51]. Furthermore, delayed intranasal delivery of hypoxic-preconditioned BMSCs significantly enhanced cell's homing to the ischemic region, optimized the therapeutic efficacy, decreased cell death in the peri-infarct region and reduced infarct volume in mice. All of these results provide promising therapeutic strategy for stroke.

7.2.3 Gene Vectors

Intranasal delivery of HMGB1 siRNA markedly reduced infarct volume in the post-ischemic rat brain (maximal reduction to $42.8 \pm 5.6\%$ at 48 h after 60 min MCAO). In addition, this protective effect was manifested by recoveries from neurological and behavioral deficits, which indicated that intranasal delivery of HMGB1 siRNA confers robust neuroprotection in the post-ischemic brain [52].

7.2.4 Small Molecules and Others

Our recent study showed that salvinorin A, a kappa opioid receptor (KOR) agonist, could potentially protect the brain and improve neurological outcome via blood brain barrier protection, apoptosis reduction and inflammation inhibition in a mouse MCAO (Middle Cerebral Artery Occlusion) model [54]. Importantly, as a non-invasive and quick method for treatment compared with other methods such as the intravenous or intramuscular injection, we demonstrated that intranasally applied molecules may bypass the blood–brain barrier and the viability of the nasal pathway to the CNS along olfactory or trigeminal associated extracellular pathways. A total volume of 10 μL of the SA solution (25% DMSO as the solvent) was administered intranasally over approximately 5 min using pipette tips. The tip was inserted into each naris so that the solution was administered to the upper nasal passage in the general area of the olfactory epithelium. However, further study was needed to show the delivery efficiency or use the absorption enhancers to improve the efficiency. Furthermore, novel drug formulations for effective brain targeting and clinical usage have to be developed.

Opioids are commonly used in critically ill patients. However, the role of opioid in modulating brain ischemia was not clear and has to be elucidated. In our study, we found that intranasal administration of Salvinorin A could reduce the infarct volume and improve the neurological outcome via inhibiting the apoptosis and the inflammation. Unlike other KOR agonists, salvinorin A does not belong to the opioid family, which produces dysphoric effects. Many intrinsic characters of the compound, i.e. naturally available from abundant plant, quick onset, lipid soluble, easy to pass blood brain barrier, sedative and antinociceptive effect, negative pathological finding in vital organs with high dose or prolonged exposure (non-toxic) and no respiratory depression, make it a potential therapeutic medication as a non-opioid KOR agonist for various neurological conditions.

Besides, in order to further improve the therapeutic effect of substances, studies have been conducted with using of gelatin nanoparticles (GNPs). One group showed that the use of GNPs as a carrier for intranasal delivery of osteopontin (OPN) peptide allowed for a 71.57% reduction in the infarct volume and extended the therapeutic window to at least 6 h post-MCAO [63]. Also, the treatment efficacy of intranasal iNOS siRNA encapsulated in GNPs delivery was investigated. Suppressed infarct volume and reductions in neurological and behavioral deficits were observed. Importantly, therapeutic potency of iNOS siRNA/GNPs was greater and sustained longer than that of bare siRNA [53]. Another group designs polycation-shielded Ca^{2+}/nucleotide nanocomplexes with simple mixing, which produce 10–25 nm sized particles. The nanocomplexes release nucleotides in response to acidic pH, which enhance cell survival rates under unfavorable conditions such as low temperature or hypoxia. Critically, the nanocomplexes reduce cerebral infarct volume in a post-ischemic rat model [55]. Previous study supported that compared with free basic fibroblast growth factor (bFGF), nano liposomal therapy was able to improve the accumulation of bFGF in brain tissues including the most affected penumbra regions which could be rescued [43].

7.3 Merits and Issues of Intranasal Administration in Stroke

Intranasal administration is a useful method to deliver drugs to the brain of ischemic stroke. We have mentioned many substances above with intranasal administration, which are potent in the treatment of ischemic stroke. Here we conclude the merits and potential issues of intranasal administration, as well as the future direction.

7.3.1 Merits

First, intranasal administration provides a direct route into the brain, which could bypass the BBB and first-past effect of liver and gastrointestinal degradation. This increases the bioavailability of drugs, especially for peptides and substances of large mass weight, if with the help of absorption enhancers [64–67]. Large surface area for absorption (human ~160 cm^2) also promotes the absorption of the drug [68].

Besides, as the drugs directly entering the brain, systemic side effects like gastrointestinal irritation will be minimized [64, 67] and the substances will stay in higher concentration (100-fold for IGF-1 [29]) with less time to reach the targeted area (about 30 min or less [67]), compared with systemic administration [69].

What's more, from clinical prospect, intranasal administration also has its own advantages. It can:

1. Expand window for thrombolytic therapy. There is only 3- or 4.5-h time window for robust thrombolytic therapy [70]. Current stroke therapy usually fails to reach patients because of delays following stroke onset, which limits the recovery of patients. However, owing to its operability, intranasal administration could be easily deployed at home or in the ambulance to extend the time window for thrombolytic treatment. Dodecafluoropentane emulsion (DDFPe) has been used to extend time window for tPA therapy in a rabbit stroke model [71]. Extended time window is crucial for the rescue of stroke, which may allow more patients more complete stroke recovery.
2. Be effective and convenient in the chronic treatment and recovery. Since peptides, proteins and other macromolecules could be transported to the brain directly through intranasal administration, substances such as PEG-TFG-alpha [21] could improve neurogenesis and the behavioral deficit in a chronic stroke model, which may be also effective in human. Additionally, as a non-invasive and painless method for treatment, intranasal administration will increase patients' compliance. When compared with other parenteral medications, its convenience makes self-medicine and quick dose adjustment possible, which is important for the treatment of chronic phase and recovery of the patients [32, 72]. Intranasal administration also reduced risk of disease transmission from application due to its non-invasiveness [68].

7.3.2 Potential Issues

Though intranasal administration is promising, potential issues still remain to be solved before its use in clinic. We will discuss the issues in the following three parts: (1) factors affecting delivery, (2) mechanism and anatomy, (3) local side effect.

7.3.2.1 Factors Affecting Delivery

Physiochemical properties of drugs, nasal environmental factors and formulation factor affect the permeation of drugs through intranasal route. Physiochemical properties of drugs include the molecular weight (MW), solubility, lipophilic-hydrophilic balance and pKa [64, 72]. Previous study observed that absorption was the highest for the compound of least molecular weight and least lipophilicity [72]. Nasal absorption also depends on the pKa of the drug and the pH of the nasal cavity, which may determine the partition of the drugs and thus affect the absorption. Modification will be needed for better and effective delivery.

Nasal environmental factors such as drug degrading enzymes [73], mucociliary clearance (MCC) [74] and pathological conditions also affect the delivery effect. The first two factors are the self-cleaning mechanisms of respiratory tract to defend inhaled heterogeneous substances, which may also prevent drugs into the nasal laminal propria. It is about 30 min before substances depositing in the nasal cavity are removed [75]. Pathological conditions such as cold and rhinitis may change the MCC and affect the permeation of drugs by hyper-secretion of nasal mucosa.

Formulation factors involve pH, viscosity, osmolarity and type of dosage form. The pH (nasal mucosa pH is 4.5–6.5 [72]) and osmolarity (close to 308 mOsmol/L [76]) of the formulation should be modified well to improve the permeation and reduce mucosal irritation. High viscosity may increase the permeation time for drugs but may also interfere with the function of normal mucosa. Dosage form may vary as different usage with different excipients to enhance absorption. Therefore, better design should be considered for the usage and purpose of drugs.

7.3.2.2 Anatomy Difference Between Rodents and Human

As we mentioned above, the mechanisms of substances delivering to brain have not been fully understood, which hurdles the further usage and modification of drugs. Besides, there are some anatomical differences between rodents and human [29, 46], which may be the problem for human test. The nasal cavity in rodents is easier to access than in human. Moreover, while the olfactory tissue covers over 50% of the nasal cavities lining in rodents, human olfactory tissue is restricted to 3–5% [77]. Therefore, drug delivery through the nasal cavity and olfactory tissue in rodents are likely to be more efficient than in human. In addition, the differences in CSF volume in rodents and human and the turnover time for CSF in these species also lead to reduced efficiency of intranasal delivery in human when compared with rodents. All of the anatomical differences might hurdle the translational research from animal to human and sophisticated models for experiment should be developed [29].

7.3.2.3 Local Side Effect

Local administration may cause local side effect, especially for prolonged medication. The histological toxicity of absorption enhancer and other excipients is not clearly demonstrated, which may be nasal irritation or disrupt the nasal membrane functions. Improper technique of administration may also deliver the drugs to other respiratory regions like lung [64], causing the loss of the dosage and irritating these regions. These restrictions will require better design of drugs and further investigation of delivery methods.

7.3.3 Future Direction

Research now mainly concentrates on the absorption. Prodrugs are utilized to get higher hydrophilic character to enhance absorption. Various prodrug formulation of L-dopa was produced and the solubility increased significantly [78]. Co-solvent, enzymatic inhibitors, muco-adhesive agents and absorption enhancer can be added in the formulation to promote the absorption by either improving permeability or reducing degradation. The use of hyaluronidase has become a routine procedure in intranasal delivery of therapeutics to increase permeability [42].

Over the last few years, novel drug formulations for effective brain targeting have been developed to improve delivery. New formulations such as liposomes, nanoparticles, microsphere and nanoemulsions [72] are very encouraging. Liposome can improve permeation of various drugs and protect them from degeneration. Nanoparticles can easily permeate to nasal lamina propria owing to their small size. Microsphere and nanoemulsions also have advantages in intranasal administration. What's more, some human experiments by intranasal administration have been done to explore the plausibility in human treatment. NeuroEPo was given to healthy volunteers by intranasal administration [79]. The side effects were well tolerated and the products may be effective in human as in rodents.

7.4 Conclusion

In brief, intranasal delivery is a potential strategy to overcome obstacles due to the BBB and is an attractive route for its non-invasiveness and quickness. Although the mechanisms involved in the delivery of different molecules from the nasal to the CNS are not yet completely understood, intranasal administration should be considered in the future for both pre-clinical and clinical studies for the treatment of ischemic stroke.

References

1. Powers WJ, et al. 2018 guidelines for the early management of patients with acute ischemic stroke: a guideline for healthcare professionals from the American Heart Association/American Stroke Association. Stroke. 2018;49(3):e46–e110.
2. Thorne RG, et al. Delivery of interferon-beta to the monkey nervous system following intranasal administration. Neuroscience. 2008;152(3):785–97.
3. Thorne RG, et al. Delivery of insulin-like growth factor-I to the rat brain and spinal cord along olfactory and trigeminal pathways following intranasal administration. Neuroscience. 2004;127(2):481–96.
4. Doty RL. The olfactory vector hypothesis of neurodegenerative disease: is it viable? Ann Neurol. 2008;63(1):7–15.

5. Kristensson K, Olsson Y. Uptake of exogenous proteins in mouse olfactory cells. Acta Neuropathol. 1971;19(2):145–54.
6. Broadwell RD, Balin BJ. Endocytic and exocytic pathways of the neuronal secretory process and trans-synaptic transfer of wheat germ agglutinin-horseradish peroxidase in vivo. J Comp Neurol. 1985;242(4):632–50.
7. Thorne RG, et al. Quantitative analysis of the olfactory pathway for drug delivery to the brain. Brain Res. 1995;692(1–2):278–82.
8. Baker H, Spencer RF. Transneuronal transport of peroxidase-conjugated wheat germ agglutinin (WGA-HRP) from the olfactory epithelium to the brain of the adult rat. Exp Brain Res. 1986;63(3):461–73.
9. Kristensson K. Microbes' roadmap to neurons. Nat Rev Neurosci. 2011;12(6):345–57.
10. Anton F, Peppel P. Central projections of trigeminal primary afferents innervating the nasal mucosa: a horseradish peroxidase study in the rat. Neuroscience. 1991;41(2–3):617–28.
11. Deatly AM, et al. Human herpes virus infections and Alzheimer's disease. Neuropathol Appl Neurobiol. 1990;16(3):213–23.
12. Jin Y, et al. Neural route of cerebral *Listeria monocytogenes* murine infection: role of immune response mechanisms in controlling bacterial neuroinvasion. Infect Immun. 2001;69(2):1093–100.
13. Balin BJ, et al. Avenues for entry of peripherally administered protein to the central nervous system in mouse, rat, and squirrel monkey. J Comp Neurol. 1986;251(2):260–80.
14. Wolburg H, et al. Epithelial and endothelial barriers in the olfactory region of the nasal cavity of the rat. Histochem Cell Biol. 2008;130(1):127–40.
15. Steinke A, et al. Molecular composition of tight and adherens junctions in the rat olfactory epithelium and fila. Histochem Cell Biol. 2008;130(2):339–61.
16. Li Y, Field PM, Raisman G. Olfactory ensheathing cells and olfactory nerve fibroblasts maintain continuous open channels for regrowth of olfactory nerve fibres. Glia. 2005;52(3):245–51.
17. Bilston LE, et al. Arterial pulsation-driven cerebrospinal fluid flow in the perivascular space: a computational model. Comput Methods Biomech Biomed Engin. 2003;6(4):235–41.
18. Schley D, et al. Mechanisms to explain the reverse perivascular transport of solutes out of the brain. J Theor Biol. 2006;238(4):962–74.
19. Wang P, Olbricht WL. Fluid mechanics in the perivascular space. J Theor Biol. 2011;274(1):52–7.
20. Hadaczek P, et al. The "perivascular pump" driven by arterial pulsation is a powerful mechanism for the distribution of therapeutic molecules within the brain. Mol Ther. 2006;14(1):69–78.
21. Guerra-Crespo M, et al. Intranasal administration of PEGylated transforming growth factor-alpha improves behavioral deficits in a chronic stroke model. J Stroke Cerebrovasc Dis. 2010;19(1):3–9.
22. Illum L. Nasal drug delivery—possibilities, problems and solutions. J Control Release. 2003;87(1–3):187–98.
23. Lee MR, et al. Oxytocin by intranasal and intravenous routes reaches the cerebrospinal fluid in rhesus macaques: determination using a novel oxytocin assay. Mol Psychiatry. 2018;23(1):115–22.
24. Rennels ML, et al. Evidence for a 'paravascular' fluid circulation in the mammalian central nervous system, provided by the rapid distribution of tracer protein throughout the brain from the subarachnoid space. Brain Res. 1985;326(1):47–63.
25. Iliff JJ, et al. A paravascular pathway facilitates CSF flow through the brain parenchyma and the clearance of interstitial solutes, including amyloid beta. Sci Transl Med. 2012;4(147):147ra111.
26. Kida S, Pantazis A, Weller RO. CSF drains directly from the subarachnoid space into nasal lymphatics in the rat. Anatomy, histology and immunological significance. Neuropathol Appl Neurobiol. 1993;19(6):480–8.
27. Szentistvanyi I, et al. Drainage of interstitial fluid from different regions of rat brain. Am J Phys. 1984;246(6 Pt 2):F835–44.

28. Liu XF, et al. The window of opportunity for treatment of focal cerebral ischemic damage with noninvasive intranasal insulin-like growth factor-I in rats. J Stroke Cerebrovasc Dis. 2004;13(1):16–23.
29. Lioutas VA, et al. Intranasal insulin and insulin-like growth factor 1 as neuroprotectants in acute ischemic stroke. Transl Stroke Res. 2015;6(4):264–75.
30. Fletcher L, et al. Intranasal delivery of erythropoietin plus insulin-like growth factor-I for acute neuroprotection in stroke. Laboratory investigation. J Neurosurg. 2009;111(1):164–70.
31. Sun BL, et al. Intranasal delivery of granulocyte colony-stimulating factor enhances its neuroprotective effects against ischemic brain injury in rats. Mol Neurobiol. 2016;53(1):320–30.
32. Zhang H, et al. Intranasal delivery of exendin-4 confers neuroprotective effect against cerebral ischemia in mice. AAPS J. 2016;18(2):385–94.
33. Kim ID, et al. Intranasal delivery of HMGB1-binding heptamer peptide confers a robust neuroprotection in the postischemic brain. Neurosci Lett. 2012;525(2):179–83.
34. Meng Y, et al. Subacute intranasal administration of tissue plasminogen activator promotes neuroplasticity and improves functional recovery following traumatic brain injury in rats. PLoS One. 2014;9(9):e106238.
35. Liu Z, et al. Subacute intranasal administration of tissue plasminogen activator increases functional recovery and axonal remodeling after stroke in rats. Neurobiol Dis. 2012;45(2):804–9.
36. Merelli A, et al. Experimental evidence of the potential use of erythropoietin by intranasal administration as a neuroprotective agent in cerebral hypoxia. Drug Metabol Drug Interact. 2011;26(2):65–9.
37. Merelli A, et al. Recovery of motor spontaneous activity after intranasal delivery of human recombinant erythropoietin in a focal brain hypoxia model induced by CoCl2 in rats. Neurotox Res. 2011;20(2):182–92.
38. Gao Y, et al. Different expression patterns of Ngb and EPOR in the cerebral cortex and hippocampus revealed distinctive therapeutic effects of intranasal delivery of Neuro-EPO for ischemic insults to the gerbil brain. J Histochem Cytochem. 2011;59(2):214–27.
39. Rodriguez Cruz Y, et al. Treatment with nasal neuro-EPO improves the neurological, cognitive, and histological state in a gerbil model of focal ischemia. ScientificWorldJournal. 2010;10:2288–300.
40. Ma M, et al. Intranasal delivery of transforming growth factor-beta1 in mice after stroke reduces infarct volume and increases neurogenesis in the subventricular zone. BMC Neurosci. 2008;9:117.
41. Wei ZZ, et al. Neuroprotective and regenerative roles of intranasal Wnt-3a administration after focal ischemic stroke in mice. J Cereb Blood Flow Metab. 2018;38(3):404–21.
42. Chen D, et al. Intranasal delivery of Apelin-13 is neuroprotective and promotes angiogenesis after ischemic stroke in mice. ASN Neuro. 2015;7(5):1759091415605114. https://doi.org/10.1177/1759091415605114.
43. Zhao HM, et al. Intranasal delivery of nerve growth factor to protect the central nervous system against acute cerebral infarction. Chin Med Sci J. 2004;19(4):257–61.
44. Yang D, et al. Taming neonatal hypoxic-ischemic brain injury by intranasal delivery of plasminogen activator inhibitor-1. Stroke. 2013;44(9):2623–7.
45. Akpan N, et al. Intranasal delivery of caspase-9 inhibitor reduces caspase-6-dependent axon/neuron loss and improves neurological function after stroke. J Neurosci. 2011;31(24):8894–904.
46. Lee JH, et al. Intranasal administration of interleukin-1 receptor antagonist in a transient focal cerebral ischemia rat model. Biomol Ther (Seoul). 2017;25(2):149–57.
47. Jin YC, et al. Intranasal delivery of RGD motif-containing osteopontin icosamer confers neuroprotection in the postischemic brain via alphavbeta3 integrin binding. Mol Neurobiol. 2016;53(8):5652–63.
48. Doyle KP, et al. Nasal administration of osteopontin peptide mimetics confers neuroprotection in stroke. J Cereb Blood Flow Metab. 2008;28(6):1235–48.

49. Wei N, et al. Delayed intranasal delivery of hypoxic-preconditioned bone marrow mesenchymal stem cells enhanced cell homing and therapeutic benefits after ischemic stroke in mice. Cell Transplant. 2013;22(6):977–91.
50. Wei ZZ, et al. Intranasal delivery of bone marrow mesenchymal stem cells improved neurovascular regeneration and rescued neuropsychiatric deficits after neonatal stroke in rats. Cell Transplant. 2015;24(3):391–402.
51. van Velthoven CT, et al. Mesenchymal stem cell transplantation attenuates brain injury after neonatal stroke. Stroke. 2013;44(5):1426–32.
52. Kim ID, et al. Intranasal delivery of HMGB1 siRNA confers target gene knockdown and robust neuroprotection in the postischemic brain. Mol Ther. 2012;20(4):829–39.
53. Kim ID, et al. Robust neuroprotective effects of intranasally delivered iNOS siRNA encapsulated in gelatin nanoparticles in the postischemic brain. Nanomedicine. 2016;12(5):1219–29.
54. Chen C, et al. The role of kappa opioid receptor in brain ischemia. Crit Care Med. 2016;44(12):e1219–25.
55. Choi YS, et al. Enhanced cell survival of pH-sensitive bioenergetic nucleotide nanoparticles in energy/oxygen-depleted cells and their intranasal delivery for reduced brain infarction. Acta Biomater. 2016;41:147–60.
56. Zhao YZ, et al. Intranasal delivery of bFGF with nanoliposomes enhances in vivo neuroprotection and neural injury recovery in a rodent stroke model. J Control Release. 2016;224:165–75.
57. Frechou M, et al. Intranasal delivery of progesterone after transient ischemic stroke decreases mortality and provides neuroprotection. Neuropharmacology. 2015;97:394–403.
58. Lu T, et al. Intranasal ginsenoside Rb1 targets the brain and ameliorates cerebral ischemia/reperfusion injury in rats. Biol Pharm Bull. 2011;34(8):1319–24.
59. Hanson LR, et al. Intranasal deferoxamine provides increased brain exposure and significant protection in rat ischemic stroke. J Pharmacol Exp Ther. 2009;330(3):679–86.
60. Dalpiaz A, et al. Brain uptake of an anti-ischemic agent by nasal administration of microparticles. J Pharm Sci. 2008;97(11):4889–903.
61. Li J, et al. Intranasal pretreatment with Z-ligustilide, the main volatile component of Rhizoma Chuanxiong, confers prophylaxis against cerebral ischemia via Nrf2 and HSP70 signaling pathways. J Agric Food Chem. 2017;65(8):1533–42.
62. Wen R, et al. Xingnaojing mPEG2000-PLA modified microemulsion for transnasal delivery: pharmacokinetic and brain-targeting evaluation. Drug Dev Ind Pharm. 2016;42(6):926–35.
63. Joachim E, et al. Gelatin nanoparticles enhance the neuroprotective effects of intranasally administered osteopontin in rat ischemic stroke model. Drug Deliv Transl Res. 2014;4(5–6):395–9.
64. Kamble MS, Bhalerao KK, Bhosale AV, Chaudhari PD. A review on nose-to-brain drug delivery. Int J Pharm Chem Sci. 2013;2(1):516–22.
65. Kiran KA. Stratergies and prospects of nasal drug delivery systems. Int J Pharm Sci Res. 2012;3(3):648–58.
66. Chien YW, Su KSE, Chang S-F. Nasal systemic drug delivery, vol. 1. New York: Marcel-Dekker; 1989. p. 1–77.
67. Bahadur S, Pathak K. Physicochemical and physiological considerations for efficient nose-to-brain targeting. Expert Opin Drug Deliv. 2012;9(1):19–31.
68. Lochhead JJ, Thorne RG. Intranasal delivery of biologics to the central nervous system. Adv Drug Deliv Rev. 2012;64(7):614–28.
69. Dogrukol-Ak D, et al. Passage of vasoactive intestinal peptide across the blood-brain barrier. Peptides. 2003;24(3):437–44.
70. Salam KA, et al. Intravenous thrombolysis for acute ischemic stroke in the 3- to 4.5-hour window—the Malabar experience. Int J Stroke. 2014;9(4):426–8.
71. Culp WC, et al. Dodecafluoropentane emulsion extends window for tPA therapy in a rabbit stroke model. Mol Neurobiol. 2015;52(2):979–84.
72. Chen Z, et al. Enhancing effect of borneol and muscone on geniposide transport across the human nasal epithelial cell monolayer. PLoS One. 2014;9(7):e101414.

73. Bhowmik D, Kharel R, Jaiswal J, Biswajit C, Kumar KP. Innovative approaches for nasal drug delivery system and its challenges and opportunities. Ann Biol Res. 2010;1(1):21–6.
74. Schipper NGM, Verhoef JC, Merkus FW. The nasal mucociliary clearance: relevance to nasal drug delivery. Pharm Res. 1991;8(7):807–14.
75. Bhumkar DR, et al. Chitosan reduced gold nanoparticles as novel carriers for transdermal delivery of insulin. Pharm Res. 2007;24:1415–27.
76. Jones N. The nose and paranasal sinuses physiology and anatomy. Adv Drug Deliv Rev. 2001;51(1–3):5–19.
77. Illum L. Is nose-to-brain transport of drugs in man a reality? J Pharm Pharmacol. 2004;56(1):3–14.
78. Kao HD, et al. Enhancement of the systemic and CNS specific delivery of L-dopa by the nasal administration of its water soluble prodrugs. Pharm Res. 2000;17(8):978–84.
79. Santos-Morales O, et al. Nasal administration of the neuroprotective candidate NeuroEPO to healthy volunteers: a randomized, parallel, open-label safety study. BMC Neurol. 2017;17(1):129.

Chapter 8
Intranasal Delivery of Drugs for Ischemic Stroke Treatment: Targeting IL-17A

Yun Lin, Jiancheng Zhang, and Jian Wang

Abstract Stroke is the second most common cause of death worldwide and a major cause of disability. However, uncertainty surrounds the efficacy and safety of peripheral or intracerebroventricular drug administration for stroke treatment. Intranasal delivery is emerging as a noninvasive option for delivering drugs to the central nervous system with minimal peripheral exposure. Use of the intranasal route could potentially reduce systemic exposure and side effects. Intranasal delivery provides rapid onset that occurs within minutes. Additionally, this method facilitates the delivery of large and/or charged molecules, which fail to effectively cross the blood-brain barrier. We have shown previously that intranasal delivery of exogenous interleukin-17A (IL-17A) promotes the survival, neuronal differentiation, and subsequent synaptogenesis of neural precursor cells in the subventricular zone during stroke recovery, as well as spontaneous recovery and angiogenesis. Therefore, although IL-17A is well-known for contributing to damage in acute ischemic stroke, it might also mediate neurorepair and spontaneous recovery after stroke when delivered intranasally.

Keywords Central nervous system · Interleukin-17 A · Intranasal delivery · Neurorepair · Stroke

---

Y. Lin
Department of Anesthesia, Institute of Anesthesia and Critical Care Medicine, Union Hospital, Tongji Medical College Huazhong University of Science and Technology, Wuhan, China

J. Zhang
Department of Critical Care Medicine, Institute of Anesthesia and Critical Care Medicine, Union Hospital, Tongji Medical College Huazhong University of Science and Technology, Wuhan, China

J. Wang (✉)
Department of Anesthesiology and Critical Care Medicine, The Johns Hopkins University School of Medicine, Baltimore, MD, USA
e-mail: jwang79@jhmi.edu

© Springer Nature Switzerland AG 2019
J. Chen et al. (eds.), *Therapeutic Intranasal Delivery for Stroke and Neurological Disorders*, Springer Series in Translational Stroke Research,
https://doi.org/10.1007/978-3-030-16715-8_8

8.1 Introduction

Stroke is the second most common cause of death and a major cause of permanent disability in adults worldwide [1, 2]. Tissue plasminogen activator (tPA) is the only drug approved by the U.S. Food and Drug Administration for thrombolytic therapy after ischemic stroke, but its efficacy and safety are limited by its narrow treatment time window and side effects [3]. In contrast, a broader window exists to promote repair and decrease stroke-associated disability in late phases. Under physiologic conditions, the normal adult brain contains two neurogenic regions: the subventricular zone (SVZ) of the lateral ventricle and the dentate gyrus of the hippocampus [4]. Although stroke induces neurogenesis in the SVZ and the migration of neural precursor cells into the injured striatum [5, 6], the contribution of endogenous neurogenesis to spontaneous recovery after stroke is exceptionally limited, leaving the affected individual with life-long neurologic deficits [7]. Angiogenesis has been shown to be coupled with neurogenesis in brain tissue repair and remodeling after stroke [8]. Therefore, therapeutic interventions are required to promote recovery after stroke by increasing SVZ neurogenesis and angiogenesis.

Traditionally, neurologic disorders, like many bodily disorders, have been treated through peripheral drug administration (predominantly oral administration). However, a variety of disadvantages are associated with using peripheral administration to treat central nervous system (CNS) diseases. Systemic administration can lead to side effects and low bioavailability as a result of first-pass hepatic and intestinal metabolism, plasma protein binding, and the ability of the blood-brain barrier (BBB) to severely restrict entry of all but small, nonpolar compounds. Substantial evidence has shown that intranasal administration represents the most promising, novel, noninvasive method for delivering therapeutic substances directly to the CNS. Here we discuss the advantages of using the intranasal route over peripheral or intracerebroventricular (ICV) routes for treating ischemic stroke. We also introduce our recent study, which showed that intranasal delivery of interleukin-17A (IL-17A) promotes neurogenesis and functional recovery in the later phases of stroke.

8.2 Intranasal Delivery for the Treatment of Neurologic Disorders

The intranasal delivery method was first developed by Frey in 1989 for targeting neurotrophic factors (e.g., nerve growth factor and fibroblast growth factor-2) to the CNS [8]. This noninvasive delivery method targets therapeutics to the CNS, reducing systemic exposure and side effects. Thus, the intranasal route can be advantageous for delivery of many CNS drugs, including those that can cross the BBB upon systemic administration. Intranasal delivery of therapeutics to the CNS is rapid, occurring within minutes.

Routing drugs directly from the nasal cavity to the brain sidesteps the first-pass effect, during which metabolism in the liver, kidney filtration, and degradation greatly reduce the amount of active drug that eventually reaches the brain [9]. The duration and intensity of a drug's actions can also be affected by the degree to which it binds to proteins within blood plasma. The more drug that binds to protein, the less efficiently it can transport across the BBB [10]. Another concern with systemic administration is adverse systemic or even toxic side effects. Although ICV injection can deliver drugs directly to the brain, it is a highly invasive and risky procedure [11]. Insufflation of drugs through the nose is noninvasive, associated with few complications, and directs compounds directly to the CNS [12–14]. Because the effect is often reached within 5 min, nasal administration can be used as an alternative to oral administration. A variety of growth factors, hormones, neuropeptides, and stem cells can be delivered intranasally. Even large and/or charged therapeutics, which do not effectively cross the BBB, can be delivered via the intranasal method. Therefore, this route holds promise for treatment of many CNS-related diseases, including stroke [14]. Nevertheless, each drug must be tested for effects on the nasal mucosa, sense of smell, and immune system, as drugs will likely enter the nasal-associated lymphatics and deep cervical lymph nodes.

8.3 Intranasal Administration of Growth Factor Confers Protective Effects Against Ischemic Stroke

Numerous experimental studies have shown that a wide variety of peptide and protein therapeutics delivered by the intranasal route have the potential to treat ischemic stroke. In a study by Liu et al. [15] intranasal administration of insulin-like growth factor-1 (IGF-1, MW = 7.65 kDa) significantly reduced infarct size by 54% when given at 2 h after ischemia induction and 39% when given at 4 h. It also improved neurologic function. Intranasal delivery of recombinant human erythropoietin (rHu-EPO) was shown to reduce neurologic and cognitive deficits, as well as histologic damage in gerbils exposed to experimentally induced focal cerebral ischemia [16]. Fletcher et al. [17] demonstrated that EPO (MW = 30–34 kDa) plus IGF-1 penetrated the brain more efficiently when delivered by the intranasal route than when delivered by intravenous, intraperitoneal, and or subcutaneous injections. The intranasal combination of EPO and IGF-1 delivered 1 h after middle cerebral artery occlusion (MCAO) significantly reduced infarct volumes 24 h later and improved neurologic function up to 90 days later. Intranasal nerve growth factor (MW = 26.5 kDa) enhanced neurogenesis in the striatum and improved functional recovery when administered 24 h after MCAO [18]. Intranasal delivery of recombinant human VEGF (MW = 38.2 kDa) also reduced infarct volume, improved behavioral recovery, and enhanced angiogenesis following MCAO in rats [19]. In mice subjected to MCAO, intranasal delivery of TGF-β1 (MW = 25 kDa) reduced infarct volume, increased neurogenesis in the SVZ, and improved functional recovery [20]. Ma et al. [21] reported improved neurologic function and reduced infarct volume in rats when basic fibroblast growth factor was delivered intranasally after cerebral ischemia/reperfusion. Rats that received intranasal basic fibroblast growth factor daily for 6 days beginning 1 day after MCAO also showed enhanced neurogenesis [22].

8.4 Intranasal Delivery of IL-17A Promotes Neurogenesis and Functional Recovery After Ischemic Stroke

The IL-17A family consists of several cytokines that participate in both acute and chronic inflammatory responses [23]. IL-17A is the most widely investigated cytokine of this family, and its production has been mainly attributed to T helper 17 (Th17) cells [23]. Recent studies have revealed that IL-17A is mainly produced by gamma delta ($\gamma\delta$) T cells in the acute phase of ischemic stroke [24, 25]. As a linkage between innate and adaptive immunity, IL-17A secreted from $\gamma\delta$ T cells plays detrimental roles in acute ischemic stroke [24, 26]. Evidence has shown that IFN-γ produced by CD4[+] T cells induces TNF-α production in macrophages, whereas IL-17A secreted by $\gamma\delta$ T cells triggers neutrophil recruitment to the infarcted hemisphere. The synergistic effect of TNF-α and IL-17A on astrocytes enhanced secretion of neutrophil-attracting chemokine CXCL-1 [25]. CXCL-1 binds to its receptor CXCR-2, resulting in the recruitment of neutrophils to the infarcted site, thus amplifying the inflammatory response and contributing to tissue damage [27]. Application of an IL-17A-blocking antibody after ischemic stroke induction decreases infarct size and improves neurologic outcome in the murine model. Additionally, IL-17A-positive lymphocytes were detected in postmortem brain tissue of patients who had experienced a stroke, suggesting that this aspect of the inflammatory cascade also occurs in the human brain [24]. In our previous study, we found that IL-17A from reactive astrocytes maintained and augmented the survival and neuronal differentiation of neural precursor cells in the SVZ during stroke recovery and subsequent synaptogenesis and spontaneous recovery through the p38 mitogen-activated protein kinase (MAPK)/calpain 1 signaling pathway [28]. We have also shown that pro-angiogenesis effects of IL-17A are involved in post-stroke functional recovery [29]. Therefore, although IL-17A is well-known for its damaging role in acute stroke, it might be an essential mediator for ischemia-induced neurorepair and spontaneous recovery. Our findings reveal a previously unrecognized role for IL-17A in neurogenesis, angiogenesis, and subsequent synaptogenesis and long-term functional recovery after ischemic stroke (Fig. 8.1) [28]. Importantly, these results indicate that IL-17A may have a biphasic role in different phases of ischemic stroke.

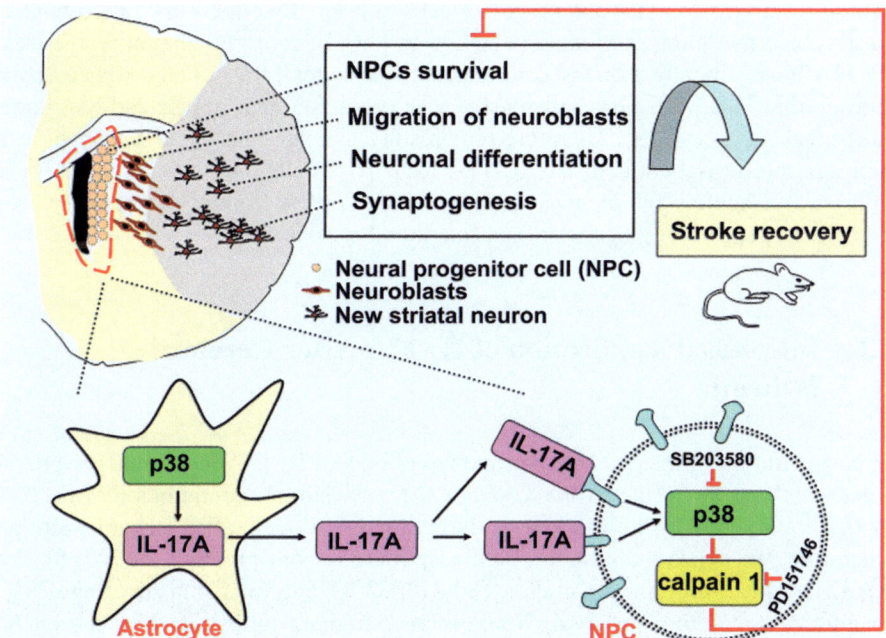

Fig. 8.1 Intranasal delivery of IL-17A promotes functional recovery by enhancing neural progenitor cell (NPC) survival, neuroblast migration, neuronal differentiation, and synaptogenesis through the p38 MAPK/calpain 1 signaling pathway

8.5 The Dual Effects of IL-17A in Different Stages of Ischemic Stroke

Signals that are deleterious during the acute stage of stroke may play beneficial roles during the recovery phase. Many reports in the literature based on cell and animal models suggest that N-methyl-D-aspartate (NMDA) receptor, the matrix metalloproteinase (MMP) family, and high-mobility-group-box-1 (HMGB1) worsen acute brain injury after stroke. However, recent studies suggest that they all could promote endogenous neurogenesis in the later phases of stroke recovery [30–33]. Another example of the biphasic nature of molecular signals involves nitric oxide (NO). Accumulating data indicate that NO is deleterious when large amounts are produced by uncontrolled neuronal or inducible nitric oxide synthase isoforms [34, 35]. Alternatively, however, NO promotes angiogenic sprouting [36]. Angiogenesis is an important feature of the peri-infarct cortex during stroke recovery [37, 38]. Therefore, some molecular signals may have biphasic roles after stroke.

Similar patterns may emerge when one looks at other stroke injury mechanisms. Among various immune cytokines, we focused on IL-17A because of two recently proposed ideas. First is the suggestion that IL-17A can exert both deleterious and beneficial effects in neuroinflammation [24, 39, 40]. Second, the crosstalk between central reactive astrocytes and precursor cells during stroke recovery supports neurovascular remodeling and functional recovery [33]. In neuroinflammatory diseases, IL-17A is specifically expressed in reactive astrocytes [40, 41]. As expected, we showed that IL-17A from reactive astrocytes promoted neurorepair and long-term functional recovery [28]. Therefore, our results indicate that IL-17A may have a biphasic role in ischemic stroke. IL-17A from γδ T cells may worsen acute brain injury in the acute stage of stroke [24], whereas IL-17A from astrocytes may promote neurogenesis, angiogenesis, and functional recovery.

8.6 Intranasal Application of IL-17A After Cerebral Ischemia

In consideration of the possible detrimental effects of IL-17A in various peripheral tissues and organs during stroke recovery, the perilesional accumulation of IL-17A in the brain seems to be the key to its neurogenic effects after ischemic stroke. Therefore, we chose the intranasal delivery route for our previous study [28]. We used a sterile 26-G Hamilton microsyringe (80330; Hamilton Company, Reno, NV) to intranasally administer 2 μL drops of recombinant mouse IL-17A (rIL-17A) diluted in PBS containing 0.1% albumin (0.1 μg/μL) or its vehicle (PBS containing 0.1% albumin) to alternating nostrils, with a 2-min interval between applications. Drops were placed at the opening of the nostril, allowing the mouse to inhale each drop into the nasal cavity. A total of 10 μL of solution, containing 1 μg rIL-17A, was delivered over the course of 5 min. The administration of rIL-17A (or vehicle) was repeated every 24 h for 2 weeks starting at 14 days post-ischemia.

We should note that, although IL-17A delivered through the nasal route may promote neurorepair and functional recovery, systemic (intravenous and intraperitoneal, etc.) administration of IL-17A may confer detrimental effects on peripheral tissues and organs because of its proinflammatory effects. Whether intranasal IL-17A can reach systemic circulation and confer detrimental effects in the peripheral system remains unknown.

8.7 Intranasal Application of EPO After Cerebral Ischemia

Another example of a compound that can be administered nasally is EPO. rHu-EPO has been tested in experimental stroke models, but its hematopoietic effect, as well as alterations in platelet function and hemostasis, might elicit potential adverse effects if used systemically in patients [42]. The main advantages of intranasal administration of EPO include the lack of hematopoietic activity and the lower doses required. Evidence has shown that intranasal delivery of rHu-EPO confers long-term neuroprotection without side effects on the hematopoietic system [43].

8.8 Conclusion

Thus, intranasal administration could be the most promising, efficient, and noninvasive route for delivering therapeutic substances directly to brain for the treatment of ischemic stroke without invasiveness or systemic adverse effects. It also could increase patient comfort and compliance. Intranasal delivery of IL-17A or other compounds may hold promise for promoting post-stroke neurovascular repair and long-term functional recovery.

References

1. Donnan GA, Fisher M, Macleod M, Davis SM. Stroke. Lancet. 2008;371:1612–23.
2. Go AS, Mozaffarian D, Roger VL, Benjamin EJ, Berry JD, Blaha MJ, Dai S, Ford ES, Fox CS, Franco S, Fullerton HJ, Gillespie C, Hailpern SM, Heit JA, Howard VJ, Huffman MD, Judd SE, Kissela BM, Kittner SJ, Lackland DT, Lichtman JH, Lisabeth LD, Mackey RH, Magid DJ, Marcus GM, Marelli A, Matchar DB, McGuire DK, Mohler ER, Moy CS, Mussolino ME, Neumar RW, Nichol G, Pandey DK, Paynter NP, Reeves MJ, Sorlie PD, Stein J, Towfighi A, Turan TN, Virani SS, Wong ND, Woo D, Turner MB. Heart disease and stroke statistics—2014 update: a report from the American Heart Association. Circulation. 2014;129:e28–e292.
3. Schwamm LH, Ali SF, Reeves MJ, Smith EE, Saver JL, Messe S, Bhatt DL, Grau-Sepulveda MV, Peterson ED, Fonarow GC. Temporal trends in patient characteristics and treatment with intravenous thrombolysis among acute ischemic stroke patients at get with the guidelines-stroke hospitals. Circ Cardiovasc Qual Outcomes. 2013;6:543–9.
4. Gross CG. Neurogenesis in the adult brain: death of a dogma. Nat Rev Neurosci. 2000;1:67–73.
5. Jin K, Wang X, Xie L, Mao XO, Zhu W, Wang Y, Shen J, Mao Y, Banwait S, Greenberg DA. Evidence for stroke-induced neurogenesis in the human brain. Proc Natl Acad Sci U S A. 2006;103:13198–202.
6. Marti-Fabregas J, Romaguera-Ros M, Gomez-Pinedo U, Martinez-Ramirez S, Jimenez-Xarrie E, Marin R, Marti-Vilalta JL, Garcia-Verdugo JM. Proliferation in the human ipsilateral subventricular zone after ischemic stroke. Neurology. 2010;74:357–65.
7. Thored P, Arvidsson A, Cacci E, Ahlenius H, Kallur T, Darsalia V, Ekdahl CT, Kokaia Z, Lindvall O. Persistent production of neurons from adult brain stem cells during recovery after stroke. Stem Cells. 2006;24:739–47.
8. Frey WH 2nd. (WO/1991/007947) neurologic agents for nasal administration to the brain (priority date 5.12.89). Geneva: World Intellectual Property Organization; 1991.
9. Bitter C, Suter-Zimmermann K, Surber C. Nasal drug delivery in humans. Curr Probl Dermatol. 2011;40:20–35.
10. Lindup WE, Orme MC. Clinical pharmacology: plasma protein binding of drugs. Br Med J (Clin Res Ed). 1981;282:212–4.
11. Jiang Y, Zhu J, Xu G, Liu X. Intranasal delivery of stem cells to the brain. Expert Opin Drug Deliv. 2011;8:623–32.
12. Dhuria SV, Hanson LR, Frey WN. Intranasal delivery to the central nervous system: mechanisms and experimental considerations. J Pharm Sci. 2010;99:1654–73.
13. Lochhead JJ, Thorne RG. Intranasal delivery of biologics to the central nervous system. Adv Drug Deliv Rev. 2012;64:614–28.
14. Chapman CD, Frey WN, Craft S, Danielyan L, Hallschmid M, Schioth HB, Benedict C. Intranasal treatment of central nervous system dysfunction in humans. Pharm Res. 2013;30:2475–84.

15. Liu XF, Fawcett JR, Hanson LR, Frey WN. The window of opportunity for treatment of focal cerebral ischemic damage with noninvasive intranasal insulin-like growth factor-I in rats. J Stroke Cerebrovasc Dis. 2004;13:16–23.
16. Rodriguez CY, Mengana TY, Munoz CA, Subiros MN, Gonzalez-Quevedo A, Sosa TI, Garcia RJ. Treatment with nasal neuro-EPO improves the neurological, cognitive, and histological state in a gerbil model of focal ischemia. ScientificWorldJournal. 2010;10:2288–300.
17. Fletcher L, Kohli S, Sprague SM, Scranton RA, Lipton SA, Parra A, Jimenez DF, Digicaylioglu M. Intranasal delivery of erythropoietin plus insulin-like growth factor-I for acute neuroprotection in stroke. Laboratory investigation. J Neurosurg. 2009;111:164–70.
18. Zhu W, Cheng S, Xu G, Ma M, Zhou Z, Liu D, Liu X. Intranasal nerve growth factor enhances striatal neurogenesis in adult rats with focal cerebral ischemia. Drug Deliv. 2011;18:338–43.
19. Yang JP, Liu HJ, Wang ZL, Cheng SM, Cheng X, Xu GL, Liu XF. The dose-effectiveness of intranasal VEGF in treatment of experimental stroke. Neurosci Lett. 2009;461:212–6.
20. Ma M, Ma Y, Yi X, Guo R, Zhu W, Fan X, Xu G, Frey WN, Liu X. Intranasal delivery of transforming growth factor-beta1 in mice after stroke reduces infarct volume and increases neurogenesis in the subventricular zone. BMC Neurosci. 2008;9:117.
21. Ma YP, Ma MM, Cheng SM, Ma HH, Yi XM, Xu GL, Liu XF. Intranasal bFGF-induced progenitor cell proliferation and neuroprotection after transient focal cerebral ischemia. Neurosci Lett. 2008;437:93–7.
22. Wang ZL, Cheng SM, Ma MM, Ma YP, Yang JP, Xu GL, Liu XF. Intranasally delivered bFGF enhances neurogenesis in adult rats following cerebral ischemia. Neurosci Lett. 2008;446:30–5.
23. Gu C, Wu L, Li X. IL-17 family: cytokines, receptors and signaling. Cytokine. 2013;64:477–85.
24. Gelderblom M, Weymar A, Bernreuther C, Velden J, Arunachalam P, Steinbach K, Orthey E, Arumugam TV, Leypoldt F, Simova O, Thom V, Friese MA, Prinz I, Holscher C, Glatzel M, Korn T, Gerloff C, Tolosa E, Magnus T. Neutralization of the IL-17 axis diminishes neutrophil invasion and protects from ischemic stroke. Blood. 2012;120:3793–802.
25. Shichita T, Sugiyama Y, Ooboshi H, Sugimori H, Nakagawa R, Takada I, Iwaki T, Okada Y, Iida M, Cua DJ, Iwakura Y, Yoshimura A. Pivotal role of cerebral interleukin-17-producing gammadeltaT cells in the delayed phase of ischemic brain injury. Nat Med. 2009;15:946–50.
26. Zhang J, Mao X, Zhou T, Cheng X, Lin Y. IL-17A contributes to brain ischemia reperfusion injury through calpain-TRPC6 pathway in mice. Neuroscience. 2014;274:419–28.
27. Veenstra M, Ransohoff RM. Chemokine receptor CXCR2: physiology regulator and neuroinflammation controller? J Neuroimmunol. 2012;246:1–9.
28. Lin Y, Zhang JC, Yao CY, Wu Y, Abdelgawad AF, Yao SL, Yuan SY. Critical role of astrocytic interleukin-17 A in post-stroke survival and neuronal differentiation of neural precursor cells in adult mice. Cell Death Dis. 2016;7:e2273.
29. Zhang J, Yao C, Chen J, Zhang Y, Yuan S, Lin Y. Hyperforin promotes post-stroke functional recovery through interleukin (IL)-17A-mediated angiogenesis. Brain Res. 2016;1646:504–13.
30. Ikonomidou C, Turski L. Why did NMDA receptor antagonists fail clinical trials for stroke and traumatic brain injury? Lancet Neurol. 2002;1:383–6.
31. Zhao BQ, Wang S, Kim HY, Storrie H, Rosen BR, Mooney DJ, Wang X, Lo EH. Role of matrix metalloproteinases in delayed cortical responses after stroke. Nat Med. 2006;12:441–5.
32. Lee SR, Kim HY, Rogowska J, Zhao BQ, Bhide P, Parent JM, Lo EH. Involvement of matrix metalloproteinase in neuroblast cell migration from the subventricular zone after stroke. J Neurosci. 2006;26:3491–5.
33. Hayakawa K, Pham LD, Katusic ZS, Arai K, Lo EH. Astrocytic high-mobility group box 1 promotes endothelial progenitor cell-mediated neurovascular remodeling during stroke recovery. Proc Natl Acad Sci U S A. 2012;109:7505–10.
34. Dirnagl U, Iadecola C, Moskowitz MA. Pathobiology of ischaemic stroke: an integrated view. Trends Neurosci. 1999;22:391–7.
35. Huang Z, Huang PL, Panahian N, Dalkara T, Fishman MC, Moskowitz MA. Effects of cerebral ischemia in mice deficient in neuronal nitric oxide synthase. Science. 1994;265:1883–5.

36. Chen J, Cui X, Zacharek A, Jiang H, Roberts C, Zhang C, Lu M, Kapke A, Feldkamp CS, Chopp M. Niaspan increases angiogenesis and improves functional recovery after stroke. Ann Neurol. 2007;62:49–58.
37. Ohab JJ, Fleming S, Blesch A, Carmichael ST. A neurovascular niche for neurogenesis after stroke. J Neurosci. 2006;26:13007–16.
38. Chopp M, Zhang ZG, Jiang Q. Neurogenesis, angiogenesis, and MRI indices of functional recovery from stroke. Stroke. 2007;38:827–31.
39. Kolls JK, Linden A. Interleukin-17 family members and inflammation. Immunity. 2004;21:467–76.
40. Hu MH, Zheng QF, Jia XZ, Li Y, Dong YC, Wang CY, Lin QY, Zhang FY, Zhao RB, Xu HW, Zhou JH, Yuan HP, Zhang WH, Ren H. Neuroprotection effect of interleukin (IL)-17 secreted by reactive astrocytes is emerged from a high-level IL-17-containing environment during acute neuroinflammation. Clin Exp Immunol. 2014;175:268–84.
41. Meng X, Zhang Y, Lao L, Saito R, Li A, Backman CM, Berman BM, Ren K, Wei PK, Zhang RX. Spinal interleukin-17 promotes thermal hyperalgesia and NMDA NR1 phosphorylation in an inflammatory pain rat model. Pain. 2013;154:294–305.
42. Hermann DM. Enhancing the delivery of erythropoietin and its variants into the ischemic brain. ScientificWorldJournal. 2009;9:967–9.
43. Merelli A, Caltana L, Girimonti P, Ramos AJ, Lazarowski A, Brusco A. Recovery of motor spontaneous activity after intranasal delivery of human recombinant erythropoietin in a focal brain hypoxia model induced by CoCl2 in rats. Neurotox Res. 2011;20:182–92.

Chapter 9
Intranasal tPA Application for Axonal Remodeling in Rodent Stroke and Traumatic Brain Injury Models

Zhongwu Liu, Ye Xiong, and Michael Chopp

Abstract Stroke and traumatic brain injury (TBI) are the major causes of adult long-term disability worldwide. Unfortunately, there are no efficacious therapies available for the vast majority of stroke and TBI patients during their convalescence. As a thrombolytic agent, recombinant tissue plasminogen activator (tPA) is the only FDA approved therapeutic agent for treatment of acute ischemic stroke; however, the application of tPA is limited by the narrow therapeutic time window and potential adverse side effects on brain edema and hemorrhage. In addition to vascular endothelium derived tPA in the circulation, neuroendocrine tissue synthesized tPA is widely distributed in the CNS and is involved in axonal path finding, synaptic plasticity and dendritic remodeling during development, and axonal outgrowth after stroke and injury. We have investigated the therapeutic effect of tPA on neurological recovery and corticospinal axonal remodeling in rodent subacute stroke and TBI models administered intranasally, to bypass the blood-brain barrier and avoid the rapid inactivation and clearance of tPA in the circulation. The neurorestorative benefits of tPA in subacute stroke and TBI treatments and the potential underlying mechanisms are discussed in this chapter.

Keywords Axonal remodeling · Functional recovery · Middle cerebral artery occlusion · Tissue plasminogen activator · Traumatic brain injury

9.1 Different Roles of tPA in the Circulation and CNS Parenchymal Tissue

Tissue-type plasminogen activator (tPA) belongs to the family of serine proteases. In the circulation, tPA is produced and released by endothelial cells [1], where it plays critical roles as a member of the fibrinolytic system to convert the zymogen plasminogen into the active protease plasmin [2, 3], which catalyzes the digestion of fibrin, the primary structural component of the blood clot, into soluble degradation products facilitating clot dissolution [4]. Recombinant tPA is the only FDA approved thrombolytic agent for the acute treatment of ischemic stroke.

By contrast, in the parenchymal tissue of the CNS, tPA is widely expressed in neurons and glial cells [5, 6], including the hippocampus [7], cerebral cortex [8], hypothalamus [9], thalamus, and cerebellum [10, 11]. Both in vitro and in vivo experiments demonstrated that tPA is localized to axon terminals [12] and dendrites [13], and involved in neurite outgrowth [14–17], neuronal migration [18, 19], synaptic plasticity [20–23], long-term potentiation (LTP) and learning [24, 25]. Thus, hippocampal-related behavioral tasks such as novel exploration, context conditioning, or spatial memory are impaired in tPA-deficient mice [26], and over-expression of tPA specifically in the hippocampus neurons leads to enhanced LTP and improved learning of tasks [27].

The roles of tPA in neuronal pathologies are controversial. Previous studies reported that tPA-deficient (tPA$^{-/-}$) male mice (SV129) exhibited smaller cerebral infarcts than wild-type (WT) male mice (C57BL/6) at 24 h after transient intravascular filament model of middle cerebral artery occlusion (MCAo, 2–3 h) [28], as well as in mice of either sex subjected to permanent ligation of the distal MCA [29]. In contrast, to avoid the variety of genetic backgrounds on vulnerability to ischemic lesion, a later study performed in tPA$^{-/-}$ versus tPA$^{+/+}$ mice on matched SV129 and C57BL/6 mixed genetic background using the similar transient filament MCAo model demonstrated that endogenous tPA protects the brain from an ischemic insult via reducing cerebrovascular fibrin deposition and the infarction volume [30]. Furthermore, in tPA$^{-/-}$ mice compared with their WT littermates with the same mixed genetic background of SV129 and C57BL/6 subjected to three different intensities of photochemical damage, the infarct size was larger in cases of mild damage, smaller in cases of severe damage, and same in cases of moderate damage [31]. In addition, although there is experimental evidence showing that intravenous administration of recombinant tPA (rtPA) to tPA$^{-/-}$ or WT mice subjected to transient MCAo produced larger infarcts than in control untreated mice [28], and intracortical delivery of the tPA inhibitor neuroserpin in rats [32], or overexpression of neuroserpin in transgenic mice [33] subjected to permanent distal MCAo reduced cerebral infarct volume and protected neurons from ischemia-induced apoptosis. However, since administration of neuroserpin also decreases the volume of ischemic lesion in tPA$^{-/-}$ mice [34], this indicates that the neuroprotective effect of neuroserpin is independent to inhibition of tPA activity. The beneficial effects of rtPA administration on reducing ischemic infarct and neuronal death are supported by multiple studies not only in rodent embolic stroke models [35–39], but also in non-thrombotic transient filament model [40, 41]. In addition, other studies demonstrate that rtPA treatment has no detectable negative effects on ischemic injury using non-thrombotic models of either transient [42] or permanent [37] cerebral ischemia.

Importantly, given the very large number of acute stroke patients worldwide who have been treated with tPA, to our knowledge there are no clinical reports indicating a neurotoxic effect of rtPA treatment. tPA-treated patients continue to improve faster and to a larger extent during the rehabilitation period beyond the acute phase than non-tPA treated patients [43]. Hippocampal neuronal death is more vulnerable in mice lacking tPA activity following an ischemic insult [44], while administration of tPA into the culture medium protects hippocampal neurons from oxygen glucose deprivation (OGD) injury [45]. Additionally, both tPA and its non-protease mutant S478A-tPA attenuated zinc toxicity in cultured cortical neurons, and when injected into cerebrospinal fluid, reduced infarcts and ameliorated motor deficits in mice subjected to permanent MCAo [46], and also reduced kainate seizure-induced hippocampal neuronal death in adult rats [47]. Following experimental stroke, endogenous tPA activity is significantly increased within the ischemic brain in rodents [28]. Overexpression of tPA specifically in neurons leads to decreasing the ischemic lesion volume and improving neurological outcome following ischemic stroke in mice [48], indicating that tPA is a neuroprotectant in the CNS. However, to determine the effect of tPA on neuronal survival, the concentration of tPA should also be considered. Some studies have reported that tPA at a high concentration greater than 200 nM (13 µg/mL) may potentiate the excitotoxic injury, while at lower concentration it could protect neurons [49, 50]. Our data also showed that tPA containing medium at a high concentration (20 µg/mL) was toxic to neurons, and at a lower concentration (7–8 µg/mL) promotes neurite growth in cultured mouse embryo cortical neurons [16, 51]. Thus, the preponderance of experimental and clinical stroke data support a neuroprotective and not a neurotoxic role for tPA.

9.2 Intranasal Delivery of tPA into the Brain

The catalytic activity of tPA is rapidly inactivated through binding of protein inhibitors, primarily plasminogen activator inhibitor-1 (PAI-1). The tPA/PAI-1 complex is cleared from the circulation by the liver. Therefore, tPA has a short half-life of 5–10 min in the bloodstream [52]. To avoid the rapid inactivation and clearance of tPA from the circulation system, we for the first time examined the effect of tPA administered by intranasal delivery on sensorimotor functional recovery in adult rodent during the subacute phase after ischemic stroke [51, 53] as well as traumatic brain injury (TBI) [54].

Intranasal delivery method has been demonstrated to directly target the brain and spinal cord along olfactory and trigeminal nerves innervating the nasal passages to bypass the blood-brain barrier [55], with a significantly lower elimination rate in the brain than intravenous administration [56]. The delivery method described by Thorne et al. [57] was modified for intranasal treatment. Briefly, under Forane anesthesia, the animals were placed in a supine position with a rolled gauze (2 × 2 in. for rats and 1 × 1 in. for mice) under the neck to maintain a horizontal head position. Saline or rtPA solution in saline (ten 6-μL drops for a total volume of 60 μL for rats and 3-μL drops for a total volume of 30 μL for mice) were placed alternately onto each nostril with a 3-min interval between drops and naturally sniffed in by the animals. The animals were kept in supine position for an additional 10 min.

To validate the efficiency of intranasal delivery, total 300 μg of recombinant human tissue plasminogen activator (rh-tPA) was administered to adult tPA knockout mice [53]. The mice were euthanized at 30 min or 120 min after the start of intranasal treatment, respectively. After transcardial perfusion with saline, the brain was removed for detection of rh-tPA delivered into the brain with a Human tPA Total Antigen ELISA Assay kit (Molecular Innovations, Novi, MI, USA). The concentrations of rh-tPA in the brain were 307 ng/mL and 228 ng/mL at 30 min and 120 min after the start of intranasal administration, respectively, suggesting the intranasal delivery is an efficacious method to deliver drugs into the brain in rodents. In contrast, tPA concentration in the extracellular space of the brain of the ischemic hemisphere and non-ischemic hemisphere in rats subjected to thromboembolic stroke model are 35 and 8 ng/mL, respectively [58].

To determine distribution of exogenous tPA in the brain, we intranasally delivered 60 μg FITC-labeled human tPA (HTPA-FITC, Molecular Innovations) to C57BL/6 mice. Animals were sacrificed 30 min later. Brains were processed for vibratome sagittal sections (100 μm). Our data demonstrate that FITC-labeled tPA was detected as early as 30 min after intranasal delivery in the olfactory bulb, and was widespread in brain regions including the cortex, striatum, subventricular zone (SVZ), DG and CA3 region of hippocampus, thalamus, cerebellum, and brain stem [51], which is in good agreement with the rapid tracer distribution study after intranasal delivery [59]. In addition, we also measured the brain tPA protein using Western blot, tPA activity with direct zymographic assay and indirect amidolytic assay, and plasmin activity by amidolytic assay in rats treated with intranasal tPA at 7 days after TBI or sham-TBI [54]. Compared to saline-treated TBI rats, both protein level and activity of tPA were significantly higher in tPA-treated sham rats and tPA-treated TBI rats 24 h after intranasal administration, while plasmin activity was higher in tPA-treated animals at 2 days after tPA treatment. Interestingly, the tPA protein level and activity in the injured brain were significantly higher than those in the sham brain although these animals received the same dose of tPA.

9.3 Intranasal tPA Administration Improves Neurological Recovery in Stroke and TBI Models

9.3.1 Intranasal Administration of tPA

tPA increases the permeability of the blood-brain barrier (BBB) in rodents [60] and humans [61], and therefore increases the risk of intracerebral hemorrhage [62]. This potential adverse side effect of tPA may aggravate brain injury and offset the therapeutic benefits when tPA is applied intravascularly early after stroke. Seven to fourteen days after the stroke is the initiation and maintenance phase of axonal sprouting response [63]. Therefore, to avoid the potential side effects on brain edema and hemorrhagic transformation during the early stage of ischemic onset, we administered tPA intranasally in the period between 7 and 14 days after stroke or TBI. Our data demonstrated that this subacute treatment does not lead to animal death or brain hemorrhage, or alter the lesion volume in either the stroke model [51, 53] or TBI model [54] employed.

To evaluate the sensorimotor functional disability and the effects of intranasal tPA treatment on neurological recovery after stroke and TBI, a series of behavioral tests were performed before surgery, 1 or 3 days after MCAo or TBI and weekly thereafter, including an adhesive-removal test [64], which measures the sensory and motor deficits by recording the time required to remove an adhesive tab from the lesion-impaired forepaw; a foot-fault test [65], which measures the accuracy of forepaw placement on a non-equidistant grid as the percentage of foot-faults of the impaired forepaw to total steps; a modified neurological severity score (mNSS) test, which is a composite of the motor (muscle status, abnormal movement), sensory (visual, tactile, and proprioceptive), and reflex tests [66]; and a single-pellet reaching test [67, 68], which characterizes the ability for skilled voluntary control of the impaired forepaw evaluated with success score of reaching food pellets.

Following intranasal administration with two doses of recombinant human tPA (600 μg/dose; Genentech Inc., San Francisco, CA) at day 7 and 14 in rat stroke and TBI models, and four doses of tPA (300 μg/dose) intranasally at day 7, 9, 11 and 13 in mouse stroke model, the sensorimotor performance of the impaired forelimb was significantly enhanced in both the unskilled tasks (i.e. adhesive-removal test, foot-fault test and mNSS) and the skilled task (i.e. single-pellet reaching test) as compared to saline-treated control animals, indicating the beneficial effects of subacute intranasal tPA treatment on neurological recovery after stroke and TBI.

In addition, intranasal tPA administration also significantly improved spatial learning evaluated using a modified Morris water maze (MWM) test [69] in rats after TBI, which is highly correlated to the number of doublecortin positive newborn neurons in the dentate gyrus of the hippocampus [54].

9.4 Intranasal tPA Administration Promotes Axonal Remodeling of the Corticospinal Tract (CST) in Stroke and TBI Models

One of the most common impairments after stroke and TBI is motor disability of the contralateral body side to the affected cerebral hemisphere. Although most patients exhibit some spontaneous behavioral improvements during the first several months after stroke or TBI, the recovery is generally incomplete. Depending on the location or the extent of damage, patients may partially recover from functional disability spontaneously with time, and recovery can be enhanced by rehabilitation training [70, 71]. Because the paralysis after stroke or TBI is a consequence of the loss or interruption of motor signals from the motor cortex to the spinal motoneurons, reestablishment of corticospinal innervation provides a physical substrate for functional recovery. The CST, the long axons of the cortical pyramidal neurons extending to the spinal cord, is the only direct descending pathway and the primary transmission tract innervating the spinal motoneurons from the sensorimotor cortex, and thus, forms the neuroanatomical basis for brain controlled voluntary movements of the peripheral muscles [72]. Clinical studies demonstrated that the extent of functional disability and the potential for functional recovery is dependent on the CST integrity in stroke patients [73–77]. Our preliminary studies have demonstrated that axonal remodeling of the CST in the spinal cord contributes to neurological recovery after stroke and TBI in rodents [68, 78–81].

Using an anterograde neuronal tracer biotinylated dextran amine (BDA, 10,000 MW; Invitrogen, Carlsbad, CA) injected into the contralesional sensorimotor cortex, we investigated the CST axonal remodeling in the denervated side of the spinal gray matter in rats subjected to stroke and TBI treated with intranasal tPA [53, 54].

At the cervical enlargement level, the BDA-positive descending CST motor fibers displayed a unilateral pattern in the spinal cord contralateral to the BDA injection side in normal animals, while few BDA-labeled CST fibers grow toward the denervated spinal gray matter in MCAo or TBI animals. In contrast, the animals that underwent MCAo or TBI and treatment with tPA showed increased BDA-positive fibers that originated from the contralesional cortical hemisphere and re-crossed the midline into the denervated side of the gray matter in the spinal cord. Furthermore, the total length of axons crossing the midline at cervical spinal cord was highly correlated to the behavioral outcome.

Additionally, intracortical microstimulation (ICMS) and electromyograms (EMG) were performed to validate the establishment of functional neuronal connections from the right intact cortex to bilateral forelimbs in rat subjected to TBI. Current thresholds were at low levels to evoke left forelimb movement, and were comparable in all normal and TBI animals (mean range 19–22 mA). In normal animals, threshold values evoking right forelimb movements were much higher than those required to evoke movement in the left forelimb (mean range 69–97 mA). However, 5 weeks after TBI the mean threshold in the contralesional right cortex eliciting movement in right-sided TBI-impaired forelimbs was significantly decreased (mean range 47–78 mA, $p < 0.05$ vs normal). Intranasal tPA treatment further reduced the mean threshold in the contralesional right cortex eliciting movement in the right-sided TBI-impaired forelimbs (mean range 45–63 mA, $p < 0.05$ vs saline controls), indicating that increased neuronal connections were established between the contralesional cortex and the impaired forelimb.

Due to the technical limitation of neural tracing using BDA injection, we are unable to label the entire CST axons. Therefore, we employed a transgenic mouse strain generated by mating Thy1-STOP-YFP strain (YFP expression is driven by neuron specific regulatory elements of the Thy1 promoter after Cre-mediated excision of STOP sequences) with Emx-Cre strain (Cre recombinase is specifically expressed in the embryonic forebrain, the area of origin of the CST), thus, in CST-YFP mice, the CST axons are completely and specifically labeled with YFP [82], to directly monitor the CST axonal change in the spinal cord under a fluorescent microscope. We measured axonal density in the central area of the cervical gray matter in CST-YFP mice subjected to right unilateral MCAo, then calculated the ratio of CST density in the impaired side to the intact side on the same sections to assess the axonal remodeling in the cervical cord. Our data showed that the axonal density in the denervated side of a stroke animal was decreased compared to a normal control. At 32 days after MCAo, there was a significant increase of CST density in the stroke-impaired side of the cervical cord of tPA-treated mice compared to the saline-treated stroke controls, and axons crossing the midline into the impaired side from the intact side were evident in the tPA-treated animals [51].

In order to identify the source of tPA treatment induced CST axonal remodeling in the denervated spinal gray matter, we injected a trans-synaptic tracer, attenuated Bartha strain of pseudorabies virus (PRV)-614 that expresses monomeric red fluorescent protein (RFP) into the stroke-impaired left forelimb muscles to retrogradely label the neural pathways from the peripheral tissue to the motor cortices in stroke mice [51]. Compared to normal mice in which the PRV-positive pyramidal neurons were primarily found in layer V of the motor areas in the right cerebral cortex, a few neurons with fluorescent labeling could be found in the symmetrical areas in the contralateral hemisphere. After stroke, the PRV labeled neurons were dramatically reduced in the ipsilesional cortex, but moderately increased in the contralesional cortex. The numbers of PRV-positive neurons were higher in both ipsilesional and contralesional cortices of stroke mice treated with intranasal tPA, compared to those administered with saline. Trans-synaptic transport of the PRV between CNS neurons occurs only at points of synaptic contact and proceeds in the retrograde direction (i.e. from postsynaptic to presynaptic neuron). Following intramuscular injection, the virus replicates in the infected muscles and infectious particles are released and taken up at synapses, thus spreading along neuronal hierarchical chains [83]. Therefore, the PRV labeled neuronal pathways between the stroke-impaired forelimb and the motor cortex are the pathways having substantial functional synaptic connections re-established by either axonal remodeling or rescued from the ischemic lesion, and not only structural axonal sprouting. Importantly, such synaptic connections must be effective to allow neuronal signal transmission for motor functional recovery after stroke. By measuring the PRV labeling within the ischemic and contralesional cortex, we are able to assess whether the tPA-induced neuronal reorganization between bilateral hemispheres contributes to functional recovery. Our data showing high correlations between functional outcome and CST density index and PRV-positive cortical neurons suggest that the effect of intranasal tPA on behavioral recovery after stroke, at least in our mouse stroke model, is attributed to CST axonal remodeling mainly originating from the ipsilesional cortex.

9.5 Mechanisms that Underlie tPA Treatment-Induced Axonal Remodeling

Although it is technically difficult to substantiate a direct causative relationship for the axonal sprouting to behavioral outcome, our studies demonstrated that the axonal remodeling is highly correlated with sensorimotor functional improvement of the forelimb after stroke and TBI treated with tPA intranasally [51, 53, 54]. To identify the molecular mechanisms underlying tPA-mediated neuroplasticity associated with functional recovery, we investigated the protein levels of brain-derived neurotrophic factor (BDNF) and its precursor form, proBDNF, in rats after TBI treated with tPA [54]. We found that intranasal tPA administration reduces proBDNF level and increases mature BDNF level in the injured brain and denervated cervical spinal cord. tPA is originally identified as a serine protease that catalyzes the conversion of the zymogen plasminogen into the active plasmin to lyse the fibrin component of a blood clot in the intravascular space [27]. BDNF is initially synthesized as a precursor proBDNF, which is proteolytically processed into mature form in the trans-Golgi network [84], or secreted and cleaved extracellularly [85, 86] by the proteolytic effect of tPA [22]. Neurotrophic factors are responsible for initiating and guiding the outgrowth of axons. BDNF binding to its receptor TrKB leads to the downstream signaling MEK/ERK activation [87, 88], which can elicit axonogenesis and CST axon growth [89–91]. Conversely, proneurotrophins often have biological effects that oppose those of mature neurotrophins [92]. proBDNF has an opposing role in neurite outgrowth to that of mature BDNF, which collapses neurite outgrowth of primary neurons [93] and negatively regulates neuronal remodeling and synaptic plasticity in the hippocampus [94].

tPA has been shown to activate matrix metalloproteinases-9 (MMP-9) through plasmin-dependent and -independent mechanisms [95], that may promote neurological recovery by modulating neurovascular remodeling, since inhibition of MMP-9 during the subacute phase (7–14 days) after stroke impairs functional recovery [96]. In addition, tPA also promotes degradation of the inhibitory proteoglycans in the extracellular matrix, which is important for synaptic remodeling and formation of new axonal varicosities [97]. Chondroitin sulfate proteoglycans (CSPGs) play a pivotal role in many neuronal growth mechanisms following injury to the spinal cord or brain [98]. tPA/plasmin degrades CSPGs including neurocan and phosphacan in the brain and promotes neurite reorganization after seizures [17]. Plasmin represents the critical enzyme that drives axonal plasticity and regeneration degrading CSPGs. tPA knockout mice exhibit attenuated neurite outgrowth and blunted sensory and motor recovery after spinal cord injury despite chondroitinase ABC (ChABC) treatment, which degrades the sugar chains of CSPGs and allows for synaptic plasticity, indicating that the tPA/plasmin cascade may act downstream of ChABC to allow for axonal plasticity improvement which enhances functional recovery after neural injury [99].

Other than its direct or plasmin-dependent proteolytic effects, tPA also has non-proteolytic effects in the CNS parenchyma. The beneficial effects of tPA may also be mediated through binding of tPA to low-density lipoprotein receptor-related protein 1 (LRP1). LRP1 is a major receptor for tPA [25]. tPA binds to LRP1 then potentiates NMDA signaling mediated by postsynaptic density protein-95, leading to the downstream signaling extracellular signal–regulated kinase (ERK) activation [87, 100], that mediates corticospinal motor axon regeneration [89]. A recent study demonstrated that exogenous rt-PA administration induces increased brain BDNF synthesis through a plasmin-independent potentiation of N-methyl-D-aspartate (NMDA) receptor signaling, while MK801, an NMDA receptor antagonist, completely abolishes the rise in mature BDNF expression induced by tPA [101]. The NMDA receptor [102, 103] and Annexin II [104] are also involved in promoting neurite outgrowth through the MEK/ERK pathway.

Recently, a pilot clinical trial has demonstrated that administration of intranasal insulin stabilized or improved cognition, function, and cerebral glucose metabolism for adults with amnestic mild cognitive impairment or Alzheimer disease [105]. The results from our studies provide an impetus for further pharmacokinetic and mechanistic studies, even future clinical trials, which may significantly impact the clinical needs of subacute and chronic treatment after stroke and TBI.

9.6 Summary

tPA has both intravascular and extravascular effects in the CNS. A wide distribution of tPA biosynthesis in the brain is associated with different actions of tPA, such as facilitating synaptic plasticity and axonal remodeling, which may contribute to neural repair. Our data demonstrated that intranasal administration of tPA at the subacute phase significantly enhances behavioral outcome after ischemic stroke and TBI, as well as increases CST axonal remodeling in the denervated side of the spinal gray matter and synaptic rewiring of the corticospinal innervation in adult rodents. Although further pharmacological studies are needed to optimize the administration dose and frequency to achieve the best outcome, the present studies provide a robust proof-of-principle that subacute treatment (7 days after stroke) of stroke and TBI with tPA delivered into the brain parenchymal tissue is neurorestorative and enhances neurological recovery.

Sources of Funding This work was supported by American Heart Association Grant 15GRNT25560025 (ZL), NIH RO1 NS062002 (YX) and R01 AG037506 (MC).

References

1. Schreiber SS, Tan Z, Sun N, Wang L, Zlokovic BV. Immunohistochemical localization of tissue plasminogen activator in vascular endothelium of stroke-prone regions of the rat brain. Neurosurgery. 1998;43:909–13.
2. Camiolo SM, Thorsen S, Astrup T. Fibrinogenolysis and fibrinolysis with tissue plasminogen activator, urokinase, streptokinase-activated human globulin, and plasmin. Proc Soc Exp Biol Med. 1971;138:277–80.
3. Vassalli JD, Sappino AP, Belin D. The plasminogen activator/plasmin system. J Clin Invest. 1991;88:1067–72.
4. Sizer IW, Wagley PF. The action of tyrosinase on thrombin, fibrinogen, and fibrin. J Biol Chem. 1951;192:213–21.
5. Sappino AP, Madani R, Huarte J, Belin D, Kiss JZ, Wohlwend A, Vassalli JD. Extracellular proteolysis in the adult murine brain. J Clin Invest. 1993;92:679–85.
6. Teesalu T, Kulla A, Simisker A, Siren V, Lawrence DA, Asser T, Vaheri A. Tissue plasminogen activator and neuroserpin are widely expressed in the human central nervous system. Thromb Haemost. 2004;92:358–68.
7. Salles FJ, Strickland S. Localization and regulation of the tissue plasminogen activator-plasmin system in the hippocampus. J Neurosci. 2002;22:2125–34.
8. Chen CC, Chu P, Brumberg JC. Experience-dependent regulation of tissue-type plasminogen activator in the mouse barrel cortex. Neurosci Lett. 2015;599:152–7.
9. Miyata S, Nakatani Y, Hayashi N, Nakashima T. Matrix-degrading enzymes tissue plasminogen activator and matrix metalloprotease-3 in the hypothalamo-neurohypophysial system. Brain Res. 2005;1058:1–9.
10. Favata MF, Horiuchi KY, Manos EJ, Daulerio AJ, Stradley DA, Feeser WS, Van Dyk DE, Pitts WJ, Earl RA, Hobbs F, Copeland RA, Magolda RL, Scherle PA, Trzaskos JM. Identification of a novel inhibitor of mitogen-activated protein kinase kinase. J Biol Chem. 1998;273:18623–32.
11. Seeds NW, Williams BL, Bickford PC. Tissue plasminogen activator induction in Purkinje neurons after cerebellar motor learning. Science. 1995;270:1992–4.
12. Shin CY, Kundel M, Wells DG. Rapid, activity-induced increase in tissue plasminogen activator is mediated by metabotropic glutamate receptor-dependent mRNA translation. J Neurosci. 2004;24:9425–33.
13. Lochner JE, Honigman LS, Grant WF, Gessford SK, Hansen AB, Silverman MA, Scalettar BA. Activity-dependent release of tissue plasminogen activator from the dendritic spines of hippocampal neurons revealed by live-cell imaging. J Neurobiol. 2006;66:564–77.
14. Minor K, Phillips J, Seeds NW. Tissue plasminogen activator promotes axonal outgrowth on CNS myelin after conditioned injury. J Neurochem. 2009;109:706–15.
15. Pittman RN, Ivins JK, Buettner HM. Neuronal plasminogen activators: cell surface binding sites and involvement in neurite outgrowth. J Neurosci. 1989;9:4269–86.
16. Qian JY, Chopp M, Liu Z. Mesenchymal stromal cells promote axonal outgrowth alone and synergistically with astrocytes via tPA. PLoS One. 2016;11:e0168345.
17. Wu YP, Siao CJ, Lu W, Sung TC, Frohman MA, Milev P, Bugge TH, Degen JL, Levine JM, Margolis RU, Tsirka SE. The tissue plasminogen activator (tPA)/plasmin extracellular proteolytic system regulates seizure-induced hippocampal mossy fiber outgrowth through a proteoglycan substrate. J Cell Biol. 2000;148:1295–304.

18. Moonen G, Grau-Wagemans MP, Selak I. Plasminogen activator-plasmin system and neuronal migration. Nature. 1982;298:753–5.
19. Seeds NW, Basham ME, Haffke SP. Neuronal migration is retarded in mice lacking the tissue plasminogen activator gene. Proc Natl Acad Sci U S A. 1999;96:14118–23.
20. Baranes D, Lederfein D, Huang YY, Chen M, Bailey CH, Kandel ER. Tissue plasminogen activator contributes to the late phase of LTP and to synaptic growth in the hippocampal mossy fiber pathway. Neuron. 1998;21:813–25.
21. Mataga N, Mizuguchi Y, Hensch TK. Experience-dependent pruning of dendritic spines in visual cortex by tissue plasminogen activator. Neuron. 2004;44:1031–41.
22. Pang PT, Teng HK, Zaitsev E, Woo NT, Sakata K, Zhen S, Teng KK, Yung WH, Hempstead BL, Lu B. Cleavage of proBDNF by tPA/plasmin is essential for long-term hippocampal plasticity. Science. 2004;306:487–91.
23. Samson AL, Medcalf RL. Tissue-type plasminogen activator: a multifaceted modulator of neurotransmission and synaptic plasticity. Neuron. 2006;50:673–8.
24. Frey U, Muller M, Kuhl D. A different form of long-lasting potentiation revealed in tissue plasminogen activator mutant mice. J Neurosci. 1996;16:2057–63.
25. Zhuo M, Holtzman DM, Li Y, Osaka H, DeMaro J, Jacquin M, Bu G. Role of tissue plasminogen activator receptor LRP in hippocampal long-term potentiation. J Neurosci. 2000;20:542–9.
26. Calabresi P, Napolitano M, Centonze D, Marfia GA, Gubellini P, Teule MA, Berretta N, Bernardi G, Frati L, Tolu M, Gulino A. Tissue plasminogen activator controls multiple forms of synaptic plasticity and memory. Eur J Neurosci. 2000;12:1002–12.
27. Madani R, Hulo S, Toni N, Madani H, Steimer T, Muller D, Vassalli JD. Enhanced hippocampal long-term potentiation and learning by increased neuronal expression of tissue-type plasminogen activator in transgenic mice. EMBO J. 1999;18:3007–12.
28. Wang YF, Tsirka SE, Strickland S, Stieg PE, Soriano SG, Lipton SA. Tissue plasminogen activator (tPA) increases neuronal damage after focal cerebral ischemia in wild-type and tPA-deficient mice. Nat Med. 1998;4:228–31.
29. Nagai N, De Mol M, Lijnen HR, Carmeliet P, Collen D. Role of plasminogen system components in focal cerebral ischemic infarction: a gene targeting and gene transfer study in mice. Circulation. 1999;99:2440–4.
30. Tabrizi P, Wang L, Seeds N, McComb JG, Yamada S, Griffin JH, Carmeliet P, Weiss MH, Zlokovic BV. Tissue plasminogen activator (tPA) deficiency exacerbates cerebrovascular fibrin deposition and brain injury in a murine stroke model: studies in tPA-deficient mice and wild-type mice on a matched genetic background. Arterioscler Thromb Vasc Biol. 1999;19:2801–6.
31. Nagai N, Zhao BQ, Suzuki Y, Ihara H, Urano T, Umemura K. Tissue-type plasminogen activator has paradoxical roles in focal cerebral ischemic injury by thrombotic middle cerebral artery occlusion with mild or severe photochemical damage in mice. J Cereb Blood Flow Metab. 2002;22:648–51.
32. Yepes M, Sandkvist M, Wong MK, Coleman TA, Smith E, Cohan SL, Lawrence DA. Neuroserpin reduces cerebral infarct volume and protects neurons from ischemia-induced apoptosis. Blood. 2000;96:569–76.
33. Cinelli P, Madani R, Tsuzuki N, Vallet P, Arras M, Zhao CN, Osterwalder T, Rulicke T, Sonderegger P. Neuroserpin, a neuroprotective factor in focal ischemic stroke. Mol Cell Neurosci. 2001;18:443–57.
34. Wu J, Echeverry R, Guzman J, Yepes M. Neuroserpin protects neurons from ischemia-induced plasmin-mediated cell death independently of tissue-type plasminogen activator inhibition. Am J Pathol. 2010;177:2576–84.
35. Bednar MM, McAuliffe T, Raymond S, Gross CE. Tissue plasminogen activator reduces brain injury in a rabbit model of thromboembolic stroke. Stroke. 1990;21:1705–9.
36. Kilic E, Hermann DM, Hossmann KA. Recombinant tissue-plasminogen activator-induced thrombolysis after cerebral thromboembolism in mice. Acta Neuropathol. 2000;99:219–22.

37. Meng W, Wang X, Asahi M, Kano T, Asahi K, Ackerman RH, Lo EH. Effects of tissue type plasminogen activator in embolic versus mechanical models of focal cerebral ischemia in rats. J Cereb Blood Flow Metab. 1999;19:1316–21.
38. Overgaard K, Sereghy T, Boysen G, Pedersen H, Diemer NH. Reduction of infarct volume and mortality by thrombolysis in a rat embolic stroke model. Stroke. 1992;23:1167–73. discussion 1174.
39. Zhang RL, Chopp M, Zhang ZG, Divine G. Early (1 h) administration of tissue plasminogen activator reduces infarct volume without increasing hemorrhagic transformation after focal cerebral embolization in rats. J Neurol Sci. 1998;160:1–8.
40. Kilic E, Hermann DM, Hossmann KA. Recombinant tissue plasminogen activator reduces infarct size after reversible thread occlusion of middle cerebral artery in mice. Neuroreport. 1999;10:107–11.
41. Kilic E, Kilic U, Bassetti CL, Hermann DM. Intravenously administered recombinant tissue-plasminogen activator attenuates neuronal injury after mild focal cerebral ischemia in mice. Neuroreport. 2004;15:687–9.
42. Klein GM, Li H, Sun P, Buchan AM. Tissue plasminogen activator does not increase neuronal damage in rat models of global and focal ischemia. Neurology. 1999;52:1381–4.
43. Meiner Z, Sajin A, Schwartz I, Tsenter J, Yovchev I, Eichel R, Ben-Hur T, Leker RR. Rehabilitation outcomes of stroke patients treated with tissue plasminogen activator. PM R. 2010;2:698–702. quiz 792.
44. Echeverry R, Wu J, Haile WB, Guzman J, Yepes M. Tissue-type plasminogen activator is a neuroprotectant in the mouse hippocampus. J Clin Invest. 2010;120:2194–205.
45. Flavin MP, Zhao G. Tissue plasminogen activator protects hippocampal neurons from oxygen-glucose deprivation injury. J Neurosci Res. 2001;63:388–94.
46. Yi JS, Kim YH, Koh JY. Infarct reduction in rats following intraventricular administration of either tissue plasminogen activator (tPA) or its non-protease mutant S478A-tPA. Exp Neurol. 2004;189:354–60.
47. Kim YH, Park JH, Hong SH, Koh JY. Nonproteolytic neuroprotection by human recombinant tissue plasminogen activator. Science. 1999;284:647–50.
48. Wu F, Wu J, Nicholson AD, Echeverry R, Haile WB, Catano M, An J, Lee AK, Duong D, Dammer EB, Seyfried NT, Tong FC, Votaw JR, Medcalf RL, Yepes M. Tissue-type plasminogen activator regulates the neuronal uptake of glucose in the ischemic brain. J Neurosci. 2012;32:9848–58.
49. Nicole O, Ali C, Docagne F, Plawinski L, MacKenzie ET, Vivien D, Buisson A. Neuroprotection mediated by glial cell line-derived neurotrophic factor: involvement of a reduction of NMDA-induced calcium influx by the mitogen-activated protein kinase pathway. J Neurosci. 2001;21:3024–33.
50. Wu F, Echeverry R, Wu J, An J, Haile WB, Cooper DS, Catano M, Yepes M. Tissue-type plasminogen activator protects neurons from excitotoxin-induced cell death via activation of the ERK1/2-CREB-ATF3 signaling pathway. Mol Cell Neurosci. 2013;52:9–19.
51. Chen N, Chopp M, Xiong Y, Qian JY, Lu M, Zhou D, He L, Liu Z. Subacute intranasal administration of tissue plasminogen activator improves stroke recovery by inducing axonal remodeling in mice. Exp Neurol. 2018;304:82–9.
52. Gravanis I, Tsirka SE. Tissue-type plasminogen activator as a therapeutic target in stroke. Expert Opin Ther Targets. 2008;12:159–70.
53. Liu Z, Li Y, Zhang L, Xin H, Cui Y, Hanson LR, Frey WH 2nd, Chopp M. Subacute intranasal administration of tissue plasminogen activator increases functional recovery and axonal remodeling after stroke in rats. Neurobiol Dis. 2012;45:804–9.
54. Meng Y, Chopp M, Zhang Y, Liu Z, An A, Mahmood A, Xiong Y. Subacute intranasal administration of tissue plasminogen activator promotes neuroplasticity and improves functional recovery following traumatic brain injury in rats. PLoS One. 2014;9:e106238.
55. Dhuria SV, Hanson LR, Frey WH 2nd. Intranasal delivery to the central nervous system: mechanisms and experimental considerations. J Pharm Sci. 2010;99:1654–73.

56. Bagger MA, Bechgaard E. The potential of nasal application for delivery to the central brain-a microdialysis study of fluorescein in rats. Eur J Pharm Sci. 2004;21:235–42.
57. Thorne RG, Pronk GJ, Padmanabhan V, Frey WH 2nd. Delivery of insulin-like growth factor-I to the rat brain and spinal cord along olfactory and trigeminal pathways following intranasal administration. Neuroscience. 2004;127:481–96.
58. Harada T, Kano T, Katayama Y, Matsuzaki T, Tejima E, Koshinaga M. Tissue plasminogen activator extravasated through the cerebral vessels: evaluation using a rat thromboembolic stroke model. Thromb Haemost. 2005;94:791–6.
59. Lochhead JJ, Wolak DJ, Pizzo ME, Thorne RG. Rapid transport within cerebral perivascular spaces underlies widespread tracer distribution in the brain after intranasal administration. J Cereb Blood Flow Metab. 2015;35:371–81.
60. Yepes M, Sandkvist M, Moore EG, Bugge TH, Strickland DK, Lawrence DA. Tissue-type plasminogen activator induces opening of the blood-brain barrier via the LDL receptor-related protein. J Clin Invest. 2003;112:1533–40.
61. Kidwell CS, Latour L, Saver JL, Alger JR, Starkman S, Duckwiler G, Jahan R, Vinuela F, Investigators UT, Kang DW, Warach S. Thrombolytic toxicity: blood brain barrier disruption in human ischemic stroke. Cerebrovasc Dis. 2008;25:338–43.
62. Hacke W, Donnan G, Fieschi C, Kaste M, von Kummer R, Broderick JP, Brott T, Frankel M, Grotta JC, Haley EC Jr, Kwiatkowski T, Levine SR, Lewandowski C, Lu M, Lyden P, Marler JR, Patel S, Tilley BC, Albers G, Bluhmki E, Wilhelm M, Hamilton S, ATLANTIS Trials Investigators, ECASS Trials Investigators, NINDS rt-PA Study Group Investigators. Association of outcome with early stroke treatment: pooled analysis of ATLANTIS, ECASS, and NINDS rt-PA stroke trials. Lancet. 2004;363:768–74.
63. Carmichael ST. Cellular and molecular mechanisms of neural repair after stroke: making waves. Ann Neurol. 2006;59:735–42.
64. Schallert T, Whishaw IQ. Bilateral cutaneous stimulation of the somatosensory system in hemidecorticate rats. Behav Neurosci. 1984;98:518–40.
65. Hernandez TD, Schallert T. Seizures and recovery from experimental brain damage. Exp Neurol. 1988;102:318–24.
66. Chen J, Li Y, Wang L, Zhang Z, Lu D, Lu M, Chopp M. Therapeutic benefit of intravenous administration of bone marrow stromal cells after cerebral ischemia in rats. Stroke. 2001;32:1005–11.
67. Farr TD, Whishaw IQ. Quantitative and qualitative impairments in skilled reaching in the mouse (Mus musculus) after a focal motor cortex stroke. Stroke. 2002;33:1869–75.
68. Liu Z, Chopp M, Ding X, Cui Y, Li Y. Axonal remodeling of the corticospinal tract in the spinal cord contributes to voluntary motor recovery after stroke in adult mice. Stroke. 2013;44:1951–6.
69. Choi SH, Woodlee MT, Hong JJ, Schallert T. A simple modification of the water maze test to enhance daily detection of spatial memory in rats and mice. J Neurosci Methods. 2006;156:182–93.
70. Mammi P, Zaccaria B, Franceschini M. Early rehabilitative treatment in patients with traumatic brain injuries: outcome at one-year follow-up. Eura Medicophys. 2006;42:17–22.
71. Rijntjes M. Mechanisms of recovery in stroke patients with hemiparesis or aphasia: new insights, old questions and the meaning of therapies. Curr Opin Neurol. 2006;19:76–83.
72. Heffner RS, Masterton RB. The role of the corticospinal tract in the evolution of human digital dexterity. Brain Behav Evol. 1983;23:165–83.
73. Caeyenberghs K, Leemans A, Geurts M, Linden CV, Smits-Engelsman BC, Sunaert S, Swinnen SP. Correlations between white matter integrity and motor function in traumatic brain injury patients. Neurorehabil Neural Repair. 2011;25:492–502.
74. Maraka S, Jiang Q, Jafari-Khouzani K, Li L, Malik S, Hamidian H, Zhang T, Lu M, Soltanian-Zadeh H, Chopp M, Mitsias PD. Degree of corticospinal tract damage correlates with motor function after stroke. Ann Clin Transl Neurol. 2014;1:891–9.
75. Ressel V, O'Gorman Tuura R, Scheer I, van Hedel HJA. Diffusion tensor imaging predicts motor outcome in children with acquired brain injury. Brain Imaging Behav. 2017;11:1373–84.

76. Schulz R, Park CH, Boudrias MH, Gerloff C, Hummel FC, Ward NS. Assessing the integrity of corticospinal pathways from primary and secondary cortical motor areas after stroke. Stroke. 2012;43:2248–51.
77. Stinear CM, Barber PA, Smale PR, Coxon JP, Fleming MK, Byblow WD. Functional potential in chronic stroke patients depends on corticospinal tract integrity. Brain. 2007;130:170–80.
78. Liu Z, Li Y, Zhang RL, Cui Y, Chopp M. Bone marrow stromal cells promote skilled motor recovery and enhance contralesional axonal connections after ischemic stroke in adult mice. Stroke. 2011;42:740–4.
79. Liu Z, Li Y, Zhang X, Savant-Bhonsale S, Chopp M. Contralesional axonal remodeling of the corticospinal system in adult rats following stroke and bone marrow stromal cell treatment. Stroke. 2008;39:2571–7.
80. Liu Z, Zhang RL, Li Y, Cui Y, Chopp M. Remodeling of the corticospinal innervation and spontaneous behavioral recovery after ischemic stroke in adult mice. Stroke. 2009;40:2546–51.
81. Zhang Y, Xiong Y, Mahmood A, Meng Y, Liu Z, Qu C, Chopp M. Sprouting of corticospinal tract axons from the contralateral hemisphere into the denervated side of the spinal cord is associated with functional recovery in adult rat after traumatic brain injury and erythropoietin treatment. Brain Res. 2010;1353:249–57.
82. Bareyre FM, Kerschensteiner M, Misgeld T, Sanes JR. Transgenic labeling of the corticospinal tract for monitoring axonal responses to spinal cord injury. Nat Med. 2005;11:1355–60.
83. Pickard GE, Smeraski CA, Tomlinson CC, Banfield BW, Kaufman J, Wilcox CL, Enquist LW, Sollars PJ. Intravitreal injection of the attenuated pseudorabies virus PRV Bartha results in infection of the hamster suprachiasmatic nucleus only by retrograde transsynaptic transport via autonomic circuits. J Neurosci. 2002;22:2701–10.
84. Seidah NG, Benjannet S, Pareek S, Chretien M, Murphy RA. Cellular processing of the neurotrophin precursors of NT3 and BDNF by the mammalian proprotein convertases. FEBS Lett. 1996;379:247–50.
85. Lee R, Kermani P, Teng KK, Hempstead BL. Regulation of cell survival by secreted proneurotrophins. Science. 2001;294:1945–8.
86. Yang J, Siao CJ, Nagappan G, Marinic T, Jing D, McGrath K, Chen ZY, Mark W, Tessarollo L, Lee FS, Lu B, Hempstead BL. Neuronal release of proBDNF. Nat Neurosci. 2009;12:113–5.
87. Martin AM, Kuhlmann C, Trossbach S, Jaeger S, Waldron E, Roebroek A, Luhmann HJ, Laatsch A, Weggen S, Lessmann V, Pietrzik CU. The functional role of the second NPXY motif of the LRP1 beta-chain in tissue-type plasminogen activator-mediated activation of N-methyl-D-aspartate receptors. J Biol Chem. 2008;283:12004–13.
88. Park SY, Lee JY, Choi JY, Park MJ, Kim DS. Nerve growth factor activates brain-derived neurotrophic factor promoter IV via extracellular signal-regulated protein kinase 1/2 in PC12 cells. Mol Cells. 2006;21:237–43.
89. Hollis ER 2nd, Jamshidi P, Low K, Blesch A, Tuszynski MH. Induction of corticospinal regeneration by lentiviral trkB-induced Erk activation. Proc Natl Acad Sci U S A. 2009;106:7215–20.
90. Tsuda Y, Kanje M, Dahlin LB. Axonal outgrowth is associated with increased ERK 1/2 activation but decreased caspase 3 linked cell death in Schwann cells after immediate nerve repair in rats. BMC Neurosci. 2011;12:12.
91. Wang Y, Yang F, Fu Y, Huang X, Wang W, Jiang X, Gritsenko MA, Zhao R, Monore ME, Pertz OC, Purvine SO, Orton DJ, Jacobs JM, Camp DG 2nd, Smith RD, Klemke RL. Spatial phosphoprotein profiling reveals a compartmentalized extracellular signal-regulated kinase switch governing neurite growth and retraction. J Biol Chem. 2011;286:18190–201.
92. Mast TG, Fadool DA. Mature and precursor brain-derived neurotrophic factor have individual roles in the mouse olfactory bulb. PLoS One. 2012;7:e31978.
93. Sun Y, Lim Y, Li F, Liu S, Lu JJ, Haberberger R, Zhong JH, Zhou XF. ProBDNF collapses neurite outgrowth of primary neurons by activating RhoA. PLoS One. 2012;7:e35883.
94. Yang J, Harte-Hargrove LC, Siao CJ, Marinic T, Clarke R, Ma Q, Jing D, Lafrancois JJ, Bath KG, Mark W, Ballon D, Lee FS, Scharfman HE, Hempstead BL. proBDNF negatively

regulates neuronal remodeling, synaptic transmission, and synaptic plasticity in hippocampus. Cell Rep. 2014;7:796–806.
95. Candelario-Jalil E, Yang Y, Rosenberg GA. Diverse roles of matrix metalloproteinases and tissue inhibitors of metalloproteinases in neuroinflammation and cerebral ischemia. Neuroscience. 2009;158:983–94.
96. Zhao BQ, Wang S, Kim HY, Storrie H, Rosen BR, Mooney DJ, Wang X, Lo EH. Role of matrix metalloproteinases in delayed cortical responses after stroke. Nat Med. 2006;12:441–5.
97. Benarroch EE. Tissue plasminogen activator: beyond thrombolysis. Neurology. 2007;69: 799–802.
98. Yi JH, Katagiri Y, Susarla B, Figge D, Symes AJ, Geller HM. Alterations in sulfated chondroitin glycosaminoglycans following controlled cortical impact injury in mice. J Comp Neurol. 2012;520:3295–313.
99. Bukhari N, Torres L, Robinson JK, Tsirka SE. Axonal regrowth after spinal cord injury via chondroitinase and the tissue plasminogen activator (tPA)/plasmin system. J Neurosci. 2011;31:14931–43.
100. Hu K, Yang J, Tanaka S, Gonias SL, Mars WM, Liu Y. Tissue-type plasminogen activator acts as a cytokine that triggers intracellular signal transduction and induces matrix metalloproteinase-9 gene expression. J Biol Chem. 2006;281:2120–7.
101. Rodier M, Prigent-Tessier A, Bejot Y, Jacquin A, Mossiat C, Marie C, Garnier P. Exogenous t-PA administration increases hippocampal mature BDNF levels. Plasmin- or NMDA-dependent mechanism? PLoS One. 2014;9:e92416.
102. Gakhar-Koppole N, Hundeshagen P, Mandl C, Weyer SW, Allinquant B, Muller U, Ciccolini F. Activity requires soluble amyloid precursor protein alpha to promote neurite outgrowth in neural stem cell-derived neurons via activation of the MAPK pathway. Eur J Neurosci. 2008;28:871–82.
103. Yasui H, Ito N, Yamamori T, Nakamura H, Okano J, Asanuma T, Nakajima T, Kuwabara M, Inanami O. Induction of neurite outgrowth by alpha-phenyl-N-tert-butylnitrone through nitric oxide release and Ras-ERK pathway in PC12 cells. Free Radic Res. 2010;44:645–54.
104. Lee HY, Hwang IY, Im H, Koh JY, Kim YH. Non-proteolytic neurotrophic effects of tissue plasminogen activator on cultured mouse cerebrocortical neurons. J Neurochem. 2007;101:1236–47.
105. Craft S, Baker LD, Montine TJ, Minoshima S, Watson GS, Claxton A, Arbuckle M, Callaghan M, Tsai E, Plymate SR, Green PS, Leverenz J, Cross D, Gerton B. Intranasal insulin therapy for Alzheimer disease and amnestic mild cognitive impairment: a pilot clinical trial. Arch Neurol. 2011;69(1):29–38.

Chapter 10
Therapeutic Intranasal Delivery for Alzheimer's Disease

Xinxin Wang and Fangxia Guan

Abstract Alzheimer's disease (AD) is an age-related detrimental neurodegenerative disorder with no effective treatment, which is clinically characterized by progressive memory decline and cognitive dysfunction, altered decision making, apraxia, language disturbances, etc., and often histologically manifested by the deposition of amyloid-beta (Aβ) plaques and the formation of neurofibrillary tangles. AD is a global health crisis, currently, more than 35 million people worldwide were estimated to be afflicted by AD, and the number is expect to increase with the aging of the society. Current therapy is based on neurotransmitter or enzyme replacement/modulation, and recently, stem cells therapy is proposed as a promising strategy for AD. However, effective strategies for AD treatment has not been achieved. One of the major problems is the blood–brain barrier (BBB), which hampers drug delivery into the brain. Intranasal (IN) route will overcome this obstacle by delivering drugs or cells directly to the central nervous system (CNS) through the olfactory and trigeminal neural pathways. Here, we demonstrate how intranasal delivery systems works and its advantages and disadvantages. Moreover, we discuss and summarize some latest findings on IN delivery of drug and cell in AD models, with a focus on the potential efficacy of treatments for AD.

Keywords Intranasal delivery · Therapy · Alzheimer's disease

X. Wang
The First Affiliated Hospital of Zhengzhou University, Zhengzhou, Henan, China

F. Guan (✉)
The First Affiliated Hospital of Zhengzhou University, Zhengzhou, Henan, China

School of Life Sciences, Zhengzhou University, Zhengzhou, Henan, China

© Springer Nature Switzerland AG 2019
J. Chen et al. (eds.), *Therapeutic Intranasal Delivery for Stroke and Neurological Disorders*, Springer Series in Translational Stroke Research, https://doi.org/10.1007/978-3-030-16715-8_10

10.1 Introduction

Alzheimer's disease (AD) is an age-dependent neurodegenerative disorder that is pathologically characterized by intracellular neurofibrillary tangles and extracellular amyloid beta (Aβ) plaque, neural apoptosis and neuron loss in the brain. Moreover, disturbance of metals homeostasis, extensive oxidative stress, mitochondrial damage and distribution, neuroinflammatory and calcium imbalance also contribute to the pathogenesis [1]. AD is the most common type of dementia and clinically characterized by progressive decline in learning and memory, aphasia, disuse, agnosia, spatial skills and executive dysfunction, as well as personality and behavior change. AD is the fifth cause of death among people over 65 years [2], its threats to life and reducing life quality of the patient and their families brings serious social and economic problems to the world. However, AD is a complex disease, the etiology and pathogenesis of AD is still unclear and effective therapeutic strategies remain unavailable.

Currently, acetylcholinesterase inhibitors (AChEIs), such as tacrine, donepezil, galantamine and rivastigmine are the main drugs for AD treatment. Besides, chelators that selectively bind to transition metals and reduce oxidative stress are also attractive approach to combat AD. In addition, nuclear factor kB (NF-kB), GSK3, peroxisome proliferator-activated receptor-g (PPAR-g) are suggested to regulate Aβ deposition, tau hyperphosphorylation and NFTs formation, oxidation, inflammation, demyelination and excitotoxicity, are potential targets for neuroprotective therapies. Despite major advances in neurotherapeutics, poor brain penetration due to the blood-brain barrier (BBB) pose a big challenge. Intranasal (IN) delivery, therefore, is emerged as a promising way since it bypasses the BBB in a noninvasive way, allowing direct drug delivery to the brain via a large surface area in the olfactory region and respiratory epithelium with less systemic side effects. In this chapter, we review IN delivery of AChEIs, natural anti-oxidants, insulin, nerve growth factor (NGF), peptides and several other molecules and the application to translational and clinical studies for AD treatment.

10.2 IN Delivery

10.2.1 Advantages and Challenges

IN delivery is a promising strategy to deliver drugs directly to the brain. Compared to oral administration, IN delivery of drugs achieves fast effects, avoids first-pass metabolism, reduces the side effects of systemic exposure, enhances practicality and compliance because it is noninvasive. However, the problems with IN delivery are mucociliary clearance of drugs and poor nasal permeability. To overcome this, mucoadhesive formulations or chemical penetration enhancers were explored and summarized in Fig. 10.1 [3]. These formula are generally safe and could enhance

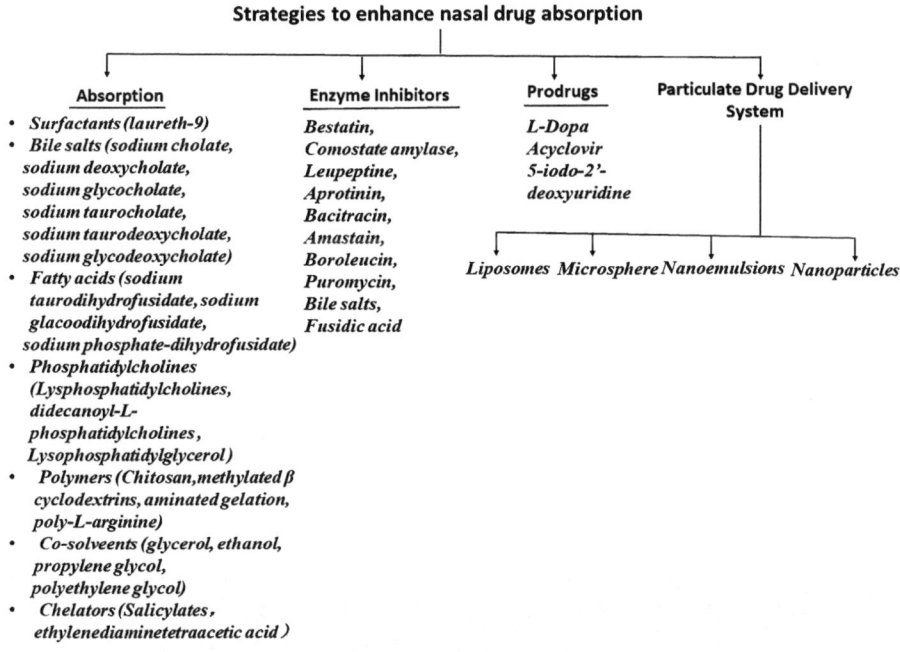

Fig. 10.1 Strategies to enhance nasal drug absorption

the stability of drugs, improve the drug absorption, protect the drugs from enzymes and chemical degradation and/or efflux back into the nasal cavity, prevent drug irritant effects, control drug release and reduce their ciliary clearance. Meanwhile, the molecular weight of polymers, free chain length, cross-link density as well as the hydration, pH, swelling, etc. should be taken into consideration for enhanced mucoadhesion.

10.2.2 Pathways of Transport from Nose to Brain

Major cerebral routes of IN delivery are olfactory pathway, rostral migratory stream pathway, and trigeminal pathway (shown in Fig. 10.2) [4].

Drugs were transported from nose to brain in intracellular or extracellular ways as shown in Fig. 10.3. The first step in intracellular transport across the olfactory and respiratory epithelia includes endocytosis into olfactory sensory neurons and trigeminal ganglion cells, respectively. This is followed by intracellular transport to olfactory bulb and brain stem, including transcytosis or transcellular transport of drug into lamina propria. Transcytosis involves the permeation of lipid soluble molecules across the apical cell membrane, intracellular space and basolateral membrane either by passive diffusion or receptor-mediated endocytosis. In terms of

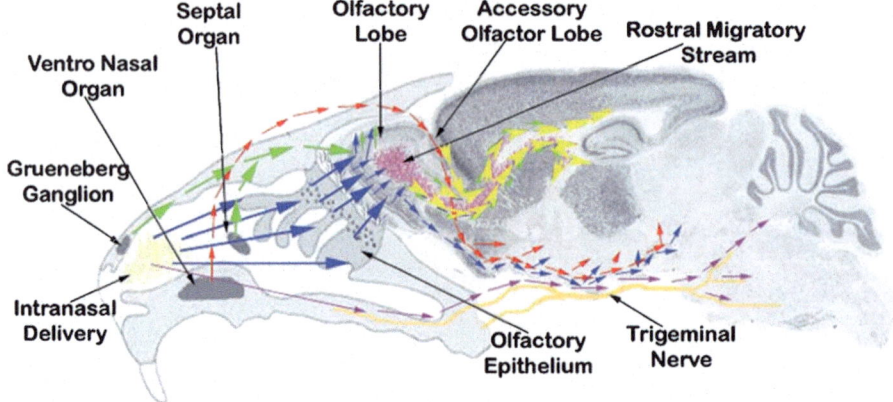

Fig. 10.2 Schema showing major routes of entry utilized after intranasal delivery of therapeutics in mice. Intranasally administered material (yellow deposits) is picked up by sensory neurons of Grueneberg ganglion, septal organ (green arrows), olfactory epithelium (blue arrow), and ventro-nasal organ (red arrow). The sensory neurons of Grueneberg ganglion, septal organ (green arrows), and olfactory epithelium (blue arrow)—all projecting to the granule cells of the olfactory lobe—eventually drain intranasally-administered material into the rostral migratory stream (RMS) (yellow arrowheads) and olfactory track at the base of the mid-brain (blue and red arrows). The material tracked into the RMS reaches the lateral and third ventricle in the close vicinity of hippocampus. The sensory neurons of ventro-nasal organ (red arrows) project to the accessory olfactory lobe, which further combine with the olfactory track at the base of the mid-brain. The material trafficked along the trigeminal nerve also combines with the olfactory track delivering to pons and hind brain, reaching to the fourth ventricle

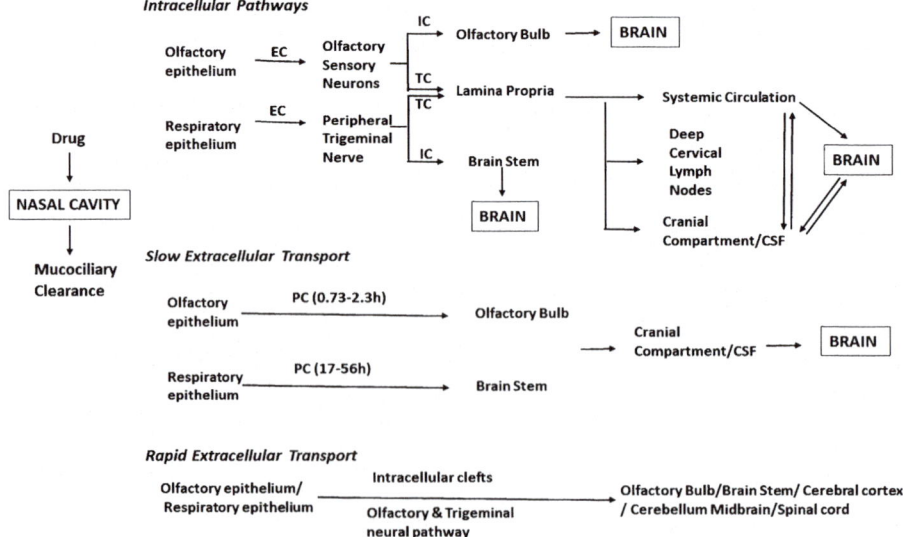

Fig. 10.3 Pathways for IN delivery system to the brain

extracellular transport, it has been estimated to take 0.73–2.3 h to diffuse from olfactory epithelium to olfactory bulb along olfactory associated extracellular pathway and 17–56 h from respiratory epithelium to brain stem along trigeminal associated extracellular pathway. This is an important pathway for the absorption of polar or hydrophilic substances, peptides and proteins. These molecules diffuse slowly from nasal membrane into the blood stream, later into the olfactory mucosa and finally transported into CNS. This pathway is less efficient with respect to transcellular pathway and is strongly dependent on drug molecular weight and size. Moreover, this mechanism is quite fast and responsible for transport of low molecular weight drugs to CNS within minutes of administration. The drugs may also be transported by rapid extracellular delivery through intercellular clefts in the olfactory and respiratory epithelium and extracellular transport along the olfactory and trigeminal neural pathway to reach the brain. Once the drug reaches lamina propria it may transport to systemic circulation; enter deep cervical lymph vessels; enter cranial compartments associated with olfactory nerve bundles.

10.3 IN Delivery Strategies for AD

IN delivery for AD treatment was first proposed by Frey in 1989. And accumulating evidence showed that IN route is a promising approach for delivery of drugs, molecules and cells in AD and is more effective than oral and intravenous (IV) route.

10.3.1 Tacrine

Tacrine (1, 2, 3, 4-tetrahydro-9-aminoacridine) is the first reversible AChEI approved for AD treatment. However, its clinical application has been limited due to low oral bioavailability, extensive hepatic first-pass effect, rapid clearance from the systemic circulation, and hepatotoxicity. To deal with these problems, Jogani et al. [5] investigated the IN delivery of tacrine, and found it could be directly transported into the brain from the nasal cavity and resulted in higher bioavailability with reduced distribution into non-targeted tissues. This selective localization of tacrine in the brain may be helpful in reducing dose, frequency of dosing and dose-dependent side effects, and proved to be an interesting new approach in delivery of the drug to the brain for the treatment of AD. Additionally, IN mucoadhesive microemulsion of tacrine improve brain targeting and fastest retrieval of memory in scopolamine-induced amnesic mice [6]. Luppi et al. reported that albumin nanoparticles carrying native and hydrophilic derivatives β-cyclodextrin derivatives can be employed for the formulation of mucoadhesive nasal formulations to modulate the mucoadhesion and permeation at the administration site [7]. Using these methods, tacrine was promising to be re-introduced for AD treatment.

10.3.2 Galantamine

Galantamine is another AChEI, however, it was discontinued for AD treatment for low aqueous solubility, dose volume limitations, and side effects such as nausea and vomiting. Therefore, researchers investigated addition of co-solvents, cyclodextrins and counter-ion exchange to enhance its solubility. Among which, galantamine-lactate represents a viable candidate for IN delivery [8]. Researchers further reported IN formulations of galantamine containing methylated-β-cyclodextrin as a stabilizer. L-a-phosphatidylcholine didecanoyl, a lipid surfactant and disodium edetate as a chelator [9] resulted in greater permeation without toxic effects to cells. In addition, galantamine hydrobromide combined with cationic chitosan nanoparticles were successfully delivered to different brain regions shortly after intranasal administration, improved pharmacological efficacy and in vivo safety, suggesting a promising way to improve AD management [10].

10.3.3 Rivastigmine

Rivastigmine is also a AChEI for AD treatment. However, the extensive first-pass metabolism and low aqueous solubility lead to poor bioactivity of the drug in vivo. Researchers found that IN administration of rivastigmine showed higher concentration in CNS regions and longer action on inhibiting the activity of AChE than intravenous (IV) administration [11]. What's more, IN administration of rivastigmine could improve distribution and pharmacological effects in CNS, especially in hippocampus, cortex and cerebrum [11]. Moreover, Shah et al. formulated rivastigmine with microemulsion (ME) and mucoadhesive microemulsions (MMEs) and found that MMEs with 0.3% w/w chitosan showed higher diffusion. Also, chitosan-modified ME are free from nasal ciliotoxicity and stable for 3 months [12]. Arumugam et al. [13]. investigated multilamellar liposomes for IN delivery of rivastigmine using soy lecithin and cholesterol by the lipid layer hydration, and showed higher AUC and Cmax compared with oral-treated group and also suggested that liposomal formulations accumulated in nasal mucosa and released the drug slowly. Fazil et al. [14] investigated IN delivery of rivastigmine loaded chitosan (CS) nanoparticles, and found the brain/blood ratio of rivastigmine was highest in the nanoparticles IN group. These results indicated that the intranasal route was a promising strategy for delivering rivastigmine and rivastigmine nanoparticles into brain.

10.3.4 Physostigmine

Physostigmine, an AChEI, is ineffective when administrated orally as it undergoes extensive first-pass metabolism. IN delivery of physostigmine combined with arecoline, a muscarinin agonist, has shown to be efficient to improve cognition. The nasal BA of physostigmine was 100% compared with IV administration and that of arecoline was 85% compared with intramuscular administration [15]. NXX-066, a physostigmine analogue, could be absorbed rapidly and completely into systemic circulation after nasal administration with Tmax of 1.5 min which was lesser than physostigmine [16]. However, the concentration of drug in CSF was very low after IN administration indicating that uptake into CSF was not enhanced by nasal administration. Therefore, the transport of drugs to CNS via IN administration may be better for poorly soluble drugs but insignificant for drugs which are completely and rapidly absorbed into systemic circulation.

10.3.5 Huperizin A

Huperizin A (Hup A), an unsaturated sesquiterpene alkaloid, is a powerful and reversible AChEI. It could easily penetrates the BBB, however, it influences peripheral cholinergic system and leads to side effects. To overcome these limitations, Zhao et al. [17] investigated nasal delivery of Hup A by means of in situ gel of gellan gum, and found that concentration of the drug after 6 h in the cerebrum, hippocampus, cerebellum, left olfactory bulb and right olfactory bulb were 1.5, 1.3, 1.0, 1.2 and 1.0 times of those after IV administration, and 2.7, 2.2, 1.9, 3.1 and 2.6 times of those after oral administration. The results revealed that IN route was a viable option for improving the brain-targeting efficiency of Hup A and also reduced the side effects to peripheral tissues. Moreover, nanoparticles have been found to improve drug transport across the epithelium due to the small particle size and the large total surface area [18].

10.3.6 Tarenflurbil

Tarenflurbil (TFB) is an Aβ42 and γ-secretase modulator. Poor brain penetration of TFB was one of the major reasons for its failure in phase III clinical trials conducted on AD patients. Thus it is urgent to improve drug delivery to brain through intranasally delivered nanocarriers. In vitro release studies proved the sustained release of TFB from nanoparticles loaded TFB (TFB-NPs and TFB-SLNs), indicating prolonged residence times of drug at targeting site. Pharmacokinetics suggested improved circulation behavior of nanoparticles and the absolute bioavailability, as well as the brain targeting efficiency. These encouraging results proved that therapeutic concentrations of TFB could be transported directly to brain via olfactory pathway after intranasal administration of polymeric and lipidic nanoparticles [19]

10.3.7 Quercetin

Quercetin, an antioxidative agent, could eliminate free radicals and protect the brain from injury. However, its therapeutic efficacy has been hampered by low solubility in the blood, rapid metabolism in the intestine and liver, and limited ability to cross the BBB. Researchers found that IN administration of quercetin liposomes modulate cognitive impairment and inhibit acetylcholinesterase activity in hippocampus of AD. This may be attributed to its antioxidant property as evidenced by decreased lipid peroxidation and increased level of antioxidant enzymes superoxide dismutase and glutathione peroxidase. Moreover, IN administration of quercetin liposomes significantly increased the survival of neurons and cholinergic neurons in hippocampus of the AD model.

10.3.8 Insulin

AD is associated with abnormal metabolism, and IV insulin administration in AD patients has been shown to improve memory recovery [20]. However, high dose is required to achieve sufficient concentration in the brain and this may lead to hypoglycemia. IN administration of insulin is a promising approach to overcome these limitations. IN administration was suggested to be safe and effective for increasing brain insulin levels, and exerts rapid effects on EEG parameters, memory, attention, mood and self-confidence without any systemic side effects [21]. IN insulin also reduced biomarker of neurodegeneration [22] and the CSF Aβ 40/42 ratio [20]. However, sex and ApoE genotype should be considered as suggested in a controlled clinical trial that only ApoE-e4-negative individuals showed significantly improvements in cognitive performance and functional abilities were relatively preserved for women [20]. In addition, glucagon-like peptide-1 (GLP-1) could stimulate insulin secretion, enhance insulin responsiveness, stimulate neuritic growth and protect against glutamate-mediated excitotoxity, oxidative stress, trophic factor withdrawal, and cell death. What's more, GLP-1 can cross BBB, and effectively reduce brain APP-Aβ burden in AD. Therefore, developing synthetic long-lasting analogues (receptor agonists) of GLP-1, e.g. Genipside or Extendin-4, can help to preserve cholinergic neuron function. Additionally, a future approach could be to genetically mesenchymal or stem cells to provide sustained delivery of neuro-stimulatory and neuro-protective agonists to restore insulin levels and functions in the brain [23].

10.3.9 Deferoxamine

Accumulation of metal leads to oxidative stress, inflammation, and contribute to neurodegenerative such as AD. Deferoxamine (DFO), a natural prototype iron chelator/radical scavenger, has been clinically applied to slow down the progression of the cognitive decline associated with iron-induced AD, however, targeting to the brain remained an issue. Hason reported that intranasal administration of DFO (2.4 mg) in C57 mice resulted in micromolar concentrations at 30 min within brain, and IN administration of 10% DFO (2.4 mg) three times a week for three in 48-week-old APP/PS1 mice significantly reduced the escape latencies in Morris water maze [24]. Guo et al. [25] reported iron-induced abnormal tau phosphorylation in cortical and hippocampal regions was suppressed by IN administration of DFO. In another study they found that IN administration of DFO reduced neuritic plaque formation, inhibited iron-induced amyloidogenic APP processing, rescued synapse loss and reversed behavioural alterations in APP/PS 1 mice [25]. And recently Fine et al. reported that IN deferoxamine affects memory loss, oxidation, and the insulin pathway in streptozotocin induced rat model of Alzheimer's disease [26].

10.3.10 R-Flurbiprofen

R-flurbiprofen was found to offer neuroprotective effects by inhibiting mitochondrial calcium overload induced by β-amyloid peptide toxicity in Alzheimer's disease (AD). However, poor brain penetration after oral administration posed a challenge to its further development for AD treatment. Study suggested that serum albumin-based nanoparticles administered via the nasal route may be a viable approach in delivering R-flurbiprofen to the brain to alleviate mitochondrial dysfunction in AD [27].

10.3.11 Curcumin

Curcumin (diferuloyl methane) has been found to exert beneficial effects on experimental models of AD by inhibiting Aβ aggregation, inflammation, tau phosphorylation in the brain, and improve memory and cognitive deficits in rats [28]. However, the poor aqueous solubility, chemical instability in alkaline medium, rapid metabolism and poor absorption from gastrointestinal tract limited its application. Chen et al. found that IN delivery of curcumin thermosensitive hydrogel resulted in short gelation time, longer mucociliary transport time and prolonged residence in nasal cavity of rats, without significant toxicity and integrity of mucocilia [29]. What's more, distribution of curcumin thermosensitive hydrogel via IN administration in cerebrum, cerebellum, hippocampus and olfactory bulb were enhanced. Some researched found that curcumin mucoadhesive nanoemulsions had a significantly higher release, higher flux and permeation across sheep nasal mucosa, with no obvious toxicity [30].

10.3.12 Piperine

Piperine (PIP) is a phytopharmaceutical with neuroprotective potential in Alzheimer's disease (AD). Oral PIP delivery is disadvantageous for the hydrophobicity and pre-systemic metabolism. Therefore, researchers developed monodisperse intranasal chitosan nanoparticles (CS-NPs) for brain targeting of PIP and found that PIP-NPs could significantly improve cognitive functions as efficient as standard drug (donpezil injection) with additional advantages of dual mechanism (Ach esterase inhibition and antioxidant effect). Meanwhile, CS-NPs could significantly alleviate PIP nasal irritation with no brain toxicity. Mucoadhesive CS-NPs were successfully tailored for effective, safe, and non-invasive PIP delivery with significant decrease in oral dose [31].

10.3.13 Angiotensin Receptor Blocker

The Renin-angiotensin system in the brain has been implicated in pathogenesis of cognitive decline. Danielyan et al. found that IN administration of losartan, an angiotensin receptor blocker, at sub-antihypertensive dose (10 mg/kg every other day for 2 months) exhibited neuroprotective effect in the APP/PS1 transgenic mouse model. There was a significantly reduction in Aβ plaques, interleukin-12, p40/p70, IL-1β, granulocytemacrophage colony-stimulating factor and increased IL-10 in mice treated with IN losartan compared with the vehicle group. The authors concluded that IN administration of losartan had direct anti-inflammatory and neuroprotective effect in CNS at concentration below than that would cause hypotensive reaction in AD patients [32].

10.3.14 Neurotrophic Factors

Neurotrophic factors plays a critical role in neural growth, regeneration and repair. IN delivery was proposed as a non-invasive technique for application of neurotrophic factors. IN delivery of NGF to the brain was rapid and efficient, and was found to decrease cholinergic deficits, phosphorylated tau and Aβ in AD11 mice [33]. Besides, some researchers found that the intranasal administration was significantly more effective than the ocular one, in rescuing the neurodegenerative phenotypic hallmarks in AD11 mice [34]. Capsoni et al. also studied the form of NGF mutated at R100 called "painless" hNGFER100 to overcome limitations of NGF due to its potent nociceptive action [35]. The mutant showed neurotrophic and anti-amyloidogenic activity in neuronal culture and a reduced nociceptive activity in vivo. Its IN administration in App X PS1 mice prevented the progress of neurodegeneration and behavioral deficits, indicating that hNGFR100 mutants variants as a new generation of therapeutics for neurodegenerative diseases.

Human acidic fibroblast growth factor (haFGF) plays significant roles in development, differentiation and regeneration of brain neurons. It regulates synaptic plasticity and processes attributed to learning and memory by improving cholinergic nerve functions [36]. However, its transport to brain is limited by BBB barrier. Lou et al. [28] investigated a novel technique of delivering haFGF14-154 to brain by fusing it with transactivator of transcription protein transduction domain, a cell penetrating peptide. And the efficacy of Tat-haFGF14-154 is markedly increased when loaded cationic liposomes for intranasal delivery in APP/PS1 mice as evidenced by ameliorated behavioral deficits, relieved brain Aβ burden, and increased the expression and activity of disintegrin and metal loproteinase domain-containing protein 10 in the brain [37].

Basic fibroblast growth factor (bFGF) promotes the survival and neurite growth of brain neurons, and modulates synaptic transmission in the hippocampus [22]. Intranasal administration of bFGF solution could help to improve the memory impairments of AD model rats, but limitations are the poor stability in nasal cavity and small transport amount. Researchers used nanoparticles conjugated with Solanum tuberosum lectin (STL), which selectively binds to N-acetylglucosamine on the nasal epithelial membrane for its brain delivery. The areas under the concentration-time curve of 125I-bFGF in the olfactory bulb, cerebrum, and cerebellum of rats following nasal application of STL modified nanoparticles (STL-bFGF-NP) were 1.79–5.17 folds of that of rats with intravenous administration, and 0.61–2.21 and 0.19–1.07 folds higher compared with intranasal solution and unmodified nanoparticles, respectively. The spatial learning and memory of AD rats in STL-bFGF-NP group were significantly better. Together with the value of choline acetyltransferase activity of rat hippocampus, the histological observations of rat hippocampal region, their study indicated that STL-NP was a promising drug delivery system for peptide and protein drugs such as bFGF to enter the CNS and play the therapeutic role.

Intranasal administration of plasma rich in growth factor PRGF Endoret to APP/PS1 mice for 4 weeks effectively reduced Aβ accumulation, tau hyperphosphorylation, astroglial activation, synaptic loss, and inflammatory responses, while promoted Aβ degradation, stimulated global improvements in anxiety, learning, and memory behaviors [38], suggesting that IN delivery of PRGF-Endoret may hold promise as an innovative therapy in AD.

10.3.15 Peptide

Vasoactive intestinal peptide (VIP) is a major neuropeptide has been found to be neuroprotective and plays important role in learning and memory. Gozes et al. synthesized a potent lipohilic analogue of VIP [stearyl-norleucine17] VIP ([St-Nle17] VIP) and found it prevented Aβ-induced cell death in rat cerebral cortical cultures with greater potency than VIP. Daily i.c.v. injections of [St-Nle17]VIP significantly improved performance of animal in Morris water maze test in animals treated with the cholinergic blocker [39]. Another study showed that daily intranasal administration of PEI-conjugated R8-Aβ(25–35) peptide significantly reduced Aβ amyloid accumulation and ameliorated the memory deficits of the transgenic mice [40]. Peptides corresponding to the NF-κB essential modifier (NEMO)-binding domain (NBD) of IκB kinase (IKK) or IκB kinase (IKK) specifically inhibit the induction of NF-κB activation without inhibiting the basal NF-κB activity. After intranasal administration, NBD peptide entered into the hippocampus, reduced hippocampal activation of NF-κB, suppressed hippocampal microglial activation, lowered the burden of Aβ in the hippocampus, attenuated apoptosis of hippocampal neurons, protected plasticity-related molecules, and improved memory and learning in 5XFAD mice [41]. IN delivery of H102 (a novel β-sheet breaker peptide) liposomes could significantly ameliorate spatial memory impairment of AD rats, increase the activities of ChAT and IDE and inhibit plaque deposition, with no toxicity on nasal mucosa [42]. Nasal administration of the β sheet breaker peptide AS 602704 was also suggested as an approach for treatment of Alzheimer's disease [27]. Taken together, these studies suggests that intranasal administration is a feasible route for peptide delivery.

10.3.16 Hormone

Melatonin, an indole amide neurohormone, has been found to protect neurons against Aβ toxicity and inhibit the progressive formation of β-sheets and amyloidfibrils, however, it has been found to have low oral BA, short biological half-life and erratic pharmacokinetic profile. Jayachandra Babu et al. [43] studied IN transport of melatonin using polymeric gel suspensions prepared with carbopol, carboxymethyl cellulose (CMC) and PEG400, and found that the concentration of melatonin in olfactory bulbs after IN administration were higher.

17β-estradiol and its brain-selective 17β-estradiol prodrug were proved to be an effective early-stage intervention in an AD mouse [44]. However, adverse peripheral effects and low estradiol water solubility were the main problems for its application. Water-soluble prodrugs, 3-N, N-dimethylamino butyl ester hydrochloride, 3-N, N-diethylamino propionyl ester hydrochloride and 3-N, N-trimethylamino butyl ester iodide, 17-N, N-dimethylamino butyl ester hydrochloride have been proposed to increase the solubility of 17β-estradiol [45]. In another preclinical study, estradiol solubility was enhanced by chitosan nanoparticles, which behaves as a bioadhesive material and binds strongly to the negatively charged mucin through electrostatic interactions, thus increasing significantly the half-time of clearance of estradiol. Moreover, the CSF concentration of estradiol following IN administration than that of IN administration [46].

Allopregnanolone (Allo), a neurosteroid, was proved to enhance neurogenesis in the hippocampus and restored learning and memory of AD mouse. However, low solubility pose a challenge for oral administration. Some researcher demonstrated that intranasal Allo increased hippocampal BrdU-labeled nuclei and PCNA protein levels in both aged wild type mice and young 3xTg AD mice [47].

10.3.17 Immunization

Vaccination with Aβ1-42 has been found to prevent Aβ accumulation and clearance of amyloid plaques [48]. Cattepoel et al. [49]. studied immunization of APP transgenic mice with single-chain variable fragment (scFv) derived from full IgG antibody raised against C-terminus of Aβ. scFv was found to enter brain after IN application and bind to amyloid plaques in cortex and hippocampus of APP transgenic mice, and inhibit Aβ fibril formation and neurotoxicity. Chronic IN administration of scFv was found to reduce congophilic amyloid angiopathy and Aβ plaques in cortex of transgenic AD mice. Another investigation confirmed that oligomeric amyloid- antibody (NU4) was able to enter the brain and maintain for 96 h post IN administration, and showed evidence of perikaryal and parenchymal uptake of NU4 in 5XFAD mouse brain, confirming the intranasal route as a non-invasive and efficient way of delivering therapeutics to the brain. In addition, this study demonstrated that intranasal delivery of NU4 antibody lowered cerebral amyloid- and improved spatial learning in 5XFAD mice [4]. Moreover, Wheat germ agglutinin enhanced cerebral uptake of antibody after intranasal administration in 5XFAD mice, resulted in greater reduction of cerebral Aβ compared to the unconjugated anti-Aβ antibody delivered intranasally in Alzheimer's 5XFAD model [50].

10.3.18 Cell-Based Therapy

Cell transplantation is a promising strategy for nervous system (CNS) disorders for the paracrine effect and multi-differential potential. However, the poor migration and homing of cells to the brain after IV delivery are the main barriers for effective treatment, IN provides a more efficient and targeted method for delivering cells to the brain than systemic administration. Moreover, IN delivery of therapeutic cells helps to avoid problems associated with surgical transplantation, such as the low survival rate of transplanted cells, limitations in cell dosage, immunological response and the impracticality of repeated surgical administration. Danielyan et al. reported that 7 days after IN delivery, MSCs were detected in the olfactory bulb (OB), cortex, amygdala, striatum, hippocampus, cerebellum, and brainstem of (Thy1)-h[A30P] αS transgenic mice. IN delivered macrophages could be detected in the OB, hippocampus, cortex, and cerebellum of 13-month-old APP/PS1 mice [51]. However, additional work is needed to determine the optimal dosage to achieve functional improvement in these mouse models. In another report, repeated intranasal delivery of soluble factors secreted by hMSCs in culture, in the absence of intravenous hMSCs injection, was also sufficient to diminish cerebral amyloidosis and neuroinflammation in the mice, suggesting that these may be used in combination or as a maintenance therapy after IV delivery of hMSCs [52].

10.4 Conclusion and Future Perspectives

AD is a multifactorial disease with complex pathogenesis. Various neuroprotective molecules, growth factors, viral vectors, and even stem cells, or other alternatives ways have been explored to intervene AD, however, the efficacy to deliver these agents to the brain was still low. IN administration bypasses the BBB and delivers a wide range of agents to the brain through olfactory, rostral migratory stream, and trigeminal routes. It provides a more effective approach to deliver drugs or cells. However, despite the progress made in area of IN delivery of drugs to brain, IN delivery for AD is still under preclinical stage for the safety and toxicity concerns. The extended contact of formulations with nasal mucosa may lead to irritation, tissue damage, epithelial/sub epithelial toxicity or ciliotoxicity and may result in environment suitable for microbial growth. In addition, IN drug formulation should be developed not to damage the primary olfactory nerves and the sense of smell. Moreover, long-term studies in animals and humans need to be carried out to confirm the effectiveness and drawbacks.

References

1. Sindi S, Mangialasche F, Kivipelto M. Advances in the prevention of Alzheimer's disease. F1000Prime Rep. 2015;7:50.
2. Hu J, Lin T, Gao Y, et al. The resveratrol trimer miyabenol C inhibits beta-secretase activity and beta-amyloid generation. PLoS One. 2015;10(1):e0115973.
3. Sood S, Jain K, Gowthamarajan K. Intranasal therapeutic strategies for management of Alzheimer's disease. J Drug Target. 2014;22(4):279–94.
4. Xiao C, Davis FJ, Chauhan BC, et al. Brain transit and ameliorative effects of intranasally delivered anti-amyloid-beta oligomer antibody in 5XFAD mice. J Alzheimers Dis. 2013;35(4):777–88.
5. Jogani VV, Shah PJ, Mishra P, et al. Nose-to-brain delivery of tacrine. J Pharm Pharmacol. 2007;59(9):1199–205.

6. Jogani VV, Shah PJ, Mishra P, et al. Intranasal mucoadhesive microemulsion of tacrine to improve brain targeting. Alzheimer Dis Assoc Disord. 2008;22(2):116–24.
7. Luppi B, Bigucci F, Corace G, et al. Albumin nanoparticles carrying cyclodextrins for nasal delivery of the anti-Alzheimer drug tacrine. Eur J Pharm Sci. 2011;44(4):559–65.
8. Leonard AK, Sileno AP, MacEvilly C, et al. Development of a novel high-concentration galantamine formulation suitable for intranasal delivery. J Pharm Sci. 2005;94(8):1736–46.
9. Leonard AK, Sileno AP, Brandt GC, et al. In vitro formulation optimization of intranasal galantamine leading to enhanced bioavailability and reduced emetic response in vivo. Int J Pharm. 2007;335(1–2):138–46.
10. Hanafy AS, Farid RM, ElGamal SS. Complexation as an approach to entrap cationic drugs into cationic nanoparticles administered intranasally for Alzheimer's disease management: preparation and detection in rat brain. Drug Dev Ind Pharm. 2015;41(12):2055–68.
11. Yang ZZ, Zhang YQ, Wu K, et al. Tissue distribution and pharmacodynamics of rivastigmine after intranasal and intravenous administration in rats. Curr Alzheimer Res. 2012;9(3):315–25.
12. Shah BM, Misra M, Shishoo CJ, et al. Nose to brain microemulsion-based drug delivery system of rivastigmine: formulation and ex-vivo characterization. Drug Deliv. 2015;22(7):918–30.
13. Arumugam K, Subramanian GS, Mallayasamy SR, et al. A study of rivastigmine liposomes for delivery into the brain through intranasal route. Acta Pharm. 2008;58(3):287–97.
14. Fazil M, Md S, Haque S, et al. Development and evaluation of rivastigmine loaded chitosan nanoparticles for brain targeting. Eur J Pharm Sci. 2012;47(1):6–15.
15. Hussain MA, Mollica JA. Intranasal absorption of physostigmine and arecoline. J Pharm Sci. 1991;80(8):750–1.
16. Dahlin M, Bjork E. Nasal administration of a physostigmine analogue (NXX-066) for Alzheimer's disease to rats. Int J Pharm. 2001;212(2):267–74.
17. Zhao Y, Yue P, Tao T, et al. Drug brain distribution following intranasal administration of Huperzine A in situ gel in rats. Acta Pharmacol Sin. 2007;28(2):273–8.
18. Illum L. Nanoparticulate systems for nasal delivery of drugs: a real improvement over simple systems? J Pharm Sci. 2007;96(3):473–83.
19. Muntimadugu E, Dhommati R, Jain A, et al. Intranasal delivery of nanoparticle encapsulated tarenflurbil: a potential brain targeting strategy for Alzheimer's disease. Eur J Pharm Sci. 2016;92:224–34.
20. Craft S, Newcomer J, Kanne S, et al. Memory improvement following induced hyperinsulinemia in Alzheimer's disease. Neurobiol Aging. 1996;17(1):123–30.
21. Benedict C, Hallschmid M, Hatke A, et al. Intranasal insulin improves memory in humans. Psychoneuroendocrinology. 2004;29(10):1326–34.
22. Zhang C, Chen J, Feng C, et al. Intranasal nanoparticles of basic fibroblast growth factor for brain delivery to treat Alzheimer's disease. Int J Pharm. 2014;461(1–2):192–202.
23. de la Monte SM. Early intranasal insulin therapy halts progression of neurodegeneration: progress in Alzheimer's disease therapeutics. Aging Health. 2012;8(1):61–4.
24. Hanson LR, Fine JM, Renner DB, et al. Intranasal delivery of deferoxamine reduces spatial memory loss in APP/PS1 mice. Drug Deliv Transl Res. 2012;2(3):160–8.
25. Guo C, Wang T, Zheng W, et al. Intranasal deferoxamine reverses iron-induced memory deficits and inhibits amyloidogenic APP processing in a transgenic mouse model of Alzheimer's disease. Neurobiol Aging. 2013;34(2):562–75.
26. Fine JM, Forsberg AC, Stroebel BM, et al. Intranasal deferoxamine affects memory loss, oxidation, and the insulin pathway in the streptozotocin rat model of Alzheimer's disease. J Neurol Sci. 2017;380:164–71.
27. Wong LR, Ho PC. Role of serum albumin as a nanoparticulate carrier for nose-to-brain delivery of R-flurbiprofen: implications for the treatment of Alzheimer's disease. J Pharm Pharmacol. 2017;70(1):59–69.
28. Lou G, Zhang Q, Xiao F, et al. Intranasal administration of TAT-haFGF((1)(4)(-)(1)(5)(4)) attenuates disease progression in a mouse model of Alzheimer's disease. Neuroscience. 2012;223:225–37.

29. Chen X, Zhi F, Jia X, et al. Enhanced brain targeting of curcumin by intranasal administration of a thermosensitive poloxamer hydrogel. J Pharm Pharmacol. 2013;65(6):807–16.
30. Sood S, Jain K, Gowthamarajan K. Optimization of curcumin nanoemulsion for intranasal delivery using design of experiment and its toxicity assessment. Colloids Surf B: Biointerfaces. 2014;113:330–7.
31. Elnaggar YSR, Etman SM, Abdelmonsif DA, et al. Intranasal piperine-loaded chitosan nanoparticles as brain-targeted therapy in Alzheimer's disease: optimization, biological efficacy, and potential toxicity. J Pharm Sci. 2015;104(10):3544–56.
32. Danielyan L, Klein R, Hanson LR, et al. Protective effects of intranasal losartan in the APP/PS1 transgenic mouse model of Alzheimer disease. Rejuvenation Res. 2010;13(2–3):195–201.
33. Chen XQ, Fawcett JR, Rahman YE, et al. Delivery of nerve growth factor to the brain via the olfactory pathway. J Alzheimer's Dis. 1998;1(1):35–44.
34. Capsoni S, Covaceuszach S, Ugolini G, et al. Delivery of NGF to the brain: intranasal versus ocular administration in anti-NGF transgenic mice. J Alzheimer's Dis. 2009;16(2):371–88.
35. Capsoni S, Marinelli S, Ceci M, et al. Intranasal "painless" human Nerve Growth Factor [corrected] slows amyloid neurodegeneration and prevents memory deficits in App X PS1 mice. PLoS One. 2012;7(5):e37555.
36. Reuss B, und Halbach OVB. Fibroblast growth factors and their receptors in the central nervous system. Cell Tissue Res. 2003;313(2):139–57.
37. Meng T, Cao Q, Lei P, et al. Tat-haFGF14-154 upregulates ADAM10 to attenuate the Alzheimer phenotype of APP/PS1 mice through the PI3K-CREB-IRE1alpha/XBP1 pathway. Mol Ther Nucleic Acids. 2017;7:439–52.
38. Anitua E, Pascual C, Antequera D, et al. Plasma rich in growth factors (PRGF-Endoret) reduces neuropathologic hallmarks and improves cognitive functions in an Alzheimer's disease mouse model. Neurobiol Aging. 2014;35(7):1582–95.
39. Gozes I, Bardea A, Reshef A, et al. Neuroprotective strategy for Alzheimer disease: intranasal administration of a fatty neuropeptide. Proc Natl Acad Sci U S A. 1996;93(1):427–32.
40. Cheng YS, Chen ZT, Liao TY, et al. An intranasally delivered peptide drug ameliorates cognitive decline in Alzheimer transgenic mice. EMBO Mol Med. 2017;9(5):703–15.
41. Rangasamy SB, Corbett GT, Roy A, et al. Intranasal delivery of NEMO-binding domain peptide prevents memory loss in a mouse model of Alzheimer's disease. J Alzheimer's Dis. 2015;47(2):385–402.
42. Zheng X, Shao X, Zhang C, et al. Intranasal H102 peptide-loaded liposomes for brain delivery to treat Alzheimer's disease. Pharm Res. 2015;32(12):3837–49.
43. Jayachandra Babu R, Dayal PP, Pawar K, et al. Nose-to-brain transport of melatonin from polymer gel suspensions: a microdialysis study in rats. J Drug Target. 2011;19(9):731–40.
44. Tschiffely AE, Schuh RA, Prokai-Tatrai K, et al. A comparative evaluation of treatments with 17beta-estradiol and its brain-selective prodrug in a double-transgenic mouse model of Alzheimer's disease. Horm Behav. 2016;83:39–44.
45. Al-Ghananeem AM, Traboulsi AA, Dittert LW, et al. Targeted brain delivery of 17 beta-estradiol via nasally administered water soluble prodrugs. AAPS PharmSciTech. 2002;3(1):E5.
46. Wang X, Chi N, Tang X. Preparation of estradiol chitosan nanoparticles for improving nasal absorption and brain targeting. Eur J Pharm Biopharm. 2008;70(3):735–40.
47. Stutzmann GE, Irwin RW, Solinsky CM, et al. Allopregnanolone preclinical acute pharmacokinetic and pharmacodynamic studies to predict tolerability and efficacy for Alzheimer's disease. PLoS One. 2015;10(6):e0128313.
48. Bard F, Cannon C, Barbour R, et al. Peripherally administered antibodies against amyloid beta-peptide enter the central nervous system and reduce pathology in a mouse model of Alzheimer disease. Nat Med. 2000;6(8):916–9.
49. Cattepoel S, Hanenberg M, Kulic L, et al. Chronic intranasal treatment with an anti-Abeta(30-42) scFv antibody ameliorates amyloid pathology in a transgenic mouse model of Alzheimer's disease. PLoS One. 2011;6(4):e18296.

50. Chauhan NB, Davis F, Xiao C. Wheat germ agglutinin enhanced cerebral uptake of anti-Abeta antibody after intranasal administration in 5XFAD mice. Vaccine. 2011;29(44):7631–7.
51. Danielyan L, Beer-Hammer S, Stolzing A, et al. Intranasal delivery of bone marrow-derived mesenchymal stem cells, macrophages, and microglia to the brain in mouse models of Alzheimer's and Parkinson's disease. Cell Transplant. 2014;23(Suppl 1):S123–39.
52. Harach T, Jammes F, Muller C, et al. Administrations of human adult ischemia-tolerant mesenchymal stem cells and factors reduce amyloid beta pathology in a mouse model of Alzheimer's disease. Neurobiol Aging. 2017;51:83–96.

Chapter 11
Intranasal Medication Delivery in Children for Brain Disorders

Gang Zhang, Myles R. McCrary, and Ling Wei

Abstract Intranasal administration is an attractive option for the delivery of many therapeutic agents especially for the treatment of central nervous system (CNS). In contrast to drugs that require delivery by peripheral injection, which requires blood brain barrier permeability of the injected drug for CNS delivery and may cause anxiety and infection, the intranasal route allows drugs to bypass the BBB due to its highly specialized nasal anatomy and the olfactory pathway. Due to its non-invasive nature and easy procedure, intranasal drug delivery is particularly suited for use in children and may be performed by medical staff or family members. This article will review the use of intranasal medications with a focus on their utility in children. We will provide an overview of the nasal anatomy and its impact on drug delivery, the side effects of drugs specific to intranasal delivery, and a list of the medications which are currently administered intranasally. The most common drug classes for intranasal delivery in pediatrics include sedatives and analgesia, drugs for seizure control, opioid antagonists, and antimigraine medications. In summary, intranasal delivery is a versatile method for drug application with a wide range of clinical utility, and especially effective in the pediatric population.

Keywords Intranasal · Pediatrics · Non-invasive · Drug delivery · Drug administration

---

G. Zhang
Department of Neurology, Children's Hospital of Nanjing Medical University, Nanjing, China

Department of Anesthesiology, Emory University School of Medicine, Atlanta, GA, USA

M. R. McCrary
Department of Anesthesiology, Emory University School of Medicine, Atlanta, GA, USA

L. Wei (✉)
Department of Anesthesiology, Emory University School of Medicine, Atlanta, GA, USA

Department of Neurology, Emory University School of Medicine, Atlanta, GA, USA
e-mail: lwei7@emory.edu

© Springer Nature Switzerland AG 2019
J. Chen et al. (eds.), *Therapeutic Intranasal Delivery for Stroke and Neurological Disorders*, Springer Series in Translational Stroke Research,
https://doi.org/10.1007/978-3-030-16715-8_11

11.1 Introduction

Traditionally, pediatric medications are delivered via oral, rectal, subcutaneous, intramuscular, intravenous, and, occasionally, intraosseous routes. There are benefits to each application. While most practical, oral medication delivery is slow in onset, difficult when patients are vomiting, and problematic when the patient's oral intake is restricted. In addition, children often refuse to swallow oral medications, potentially limiting their reliability. The rectal route may be used for young children but is less desirable for older children and adolescents. Parenteral delivery causes pain, anxiety, higher resource consumption, and the risk for contaminated needle-stick injury. Furthermore, intravenous access in children may be difficult for inexperienced providers. Intraosseous delivery is reserved for rare, serious emergencies. Intranasal delivery has garnered increasingly more attention. Intranasal delivery provides a non-injection route for pediatric clinicians. Importantly, this method is noninvasive, essentially painless, and particularly suited for children [1, 2]. The application may also be performed easily by parents and even patients themselves, so it is becoming a hot topic in pediatric medicine.

The abundant capillaries and lymphocytes on nasal mucosa facilitate drug absorption directly into the systemic circulation [3]. The digestive effect of enzymes on drugs in the nose is far less than that in the gastrointestinal tract, which must first undergo metabolism in the liver [4]. In addition, the olfactory tissue in direct contact with the central nervous system allows nasally administered drugs to be rapidly transported into the brain, which provides an effective route of administration for the central nervous system diseases in children. Nasal administration of vaccines is equally attractive due to its high efficacy and tolerance in children [5].

11.2 History and Development

Modern medical research on intranasal delivery has a history of several decades. An early study by Barash PG in 1980 characterized intranasal delivery of 10% hydrochloric acid cocaine solution, showing that the drug was rapidly absorbed and the plasma concentration peaked after 15–60 min [6]. Following this work, studies on nasal administration became increasingly common. In 1984, a "Seminar for nasal delivery route for systemic administration" was held in the United States. In 1991, the European Academic Conference on "Buccal and nasal administration as an alternative to the intravenous administration" was held in Paris. This route of administration has been further studied for modern pharmacological treatments [7].

11.3 Characteristics of Intranasal Delivery

11.3.1 The Nose: Anatomy and Function

The nasal cavity is a complex organ anatomical structure. The physiological characteristics of the nasal cavity influence processes including drug deposition, drug removal, and drug absorption. The external nose consists of paired nasal bones and upper and lower lateral cartilages. Internally, the nasal septum divides the nasal cavity into a right and left side. The nasal septum is mainly composed of cartilage and skin, so the drug absorption rate in this area is very low [8]. The lateral nasal wall consists of inferior and middle turbinates and occasionally a superior or supreme turbinate bone [9]. The opening of the sinuses is also found under the middle turbinates on the lateral nasal wall. The effective drug absorption area is in the turbinates that are rich in blood vessels. The lacrimal system drains into the nasal cavity below the anterior inferior aspect of the inferior turbinates [10].

11.3.1.1 Nasal Mucosa

The surface area of human nasal mucosa is about 150 cm^2. Epithelial microvilli of the mucosa, which are similar to small intestine villi, increase the effective area for drug absorption. The sub-epithelium of the nasal mucosa contains abundant capillaries and lymphatic capillaries which allow for rapid drug absorption into blood circulation [11]. The nasal mucosa also plays an important role as a first level defense against pathogens and allergens which enter the body via the nose. Mucous secreted by this specialized layer of cells can trap foreign pathogens as they enter the cavity [12, 13]. Under normal conditions, the sinuses produce around 1 quart of mucous per day; however, when inflamed, mucous production can increase more than two-fold [14]. The mucous also contains immune factors such as immunoglobulins including secretory IgA which can prevent bacteria adherence [15].

11.3.1.2 Nasal Mucosal Cilia

Mucociliary transport, which clears trapped foreign bodies, relies on both mucus production and ciliary function. Consequently, the cilia within the nose play a role in the airway defense system and are an important mediators of this first line of defense for the body [16]. Nasal hairs and the sticky mucous blanket of the nasal mucosa continuously help clear foreign bodies and prevent xenobiotics like allergens, pathogens, and foreign particles from reaching the lungs.

There are three distinct functional areas in the nasal cavity: the vestibular, olfactory and respiratory zones [17]. The vestibular zone serves as the first barrier against airborne particles and is sparsely vascularized. The lining in the vestibular zone is comprised of stratified squamous and keratinized epithelial cells with nasal hairs.

The olfactory area enables olfactory perception and is highly vascularized. The respiratory area has a mucous layer produced by highly specialized cells to serve as an efficient air-cleansing system [18]. Due to their rich vascularization, the olfactory and the respiratory zones serve as efficient absorption surfaces for topically applied drugs [19].

11.3.2 The Connection Between the Nasal Cavity and the Central Nervous System (CNS)

The nasal cavity consists of the nasal vestibule, respiratory region, and the olfactory region. Among these, the olfactory region partly overlies the cribriform plate and is located high in the nasal cavity [20]. The cribriform plate is a bony structure containing pores. Due to its close vicinity to the cerebrospinal fluid and direct interface with the nervous system, the olfactory region has been the focus of research interests for possible nose-to-brain delivery. There are three main routes by which drugs can be absorbed into the CNS following nasal administration: via the blood circulation in the respiratory region, through the mucosa in the olfactory region, and directly by the olfactory nerve [1]. Absorption via blood circulation in the respiratory region occurs primarily in the respiratory region. Along this pathway, the drug is absorbed into the systemic circulation from the nasal cavity, distributed to the BBB along with the blood, and passed through the CNS. The metabolism of drugs in this fashion is similar to that of intravenous injection, and the factors affecting targeting *in vivo* are basically the same [21]. In the olfactory region, drug absorption can occur directly through the olfactory mucosa, then transferred to the CSF. Finally, some intranasally delivered drugs can enter the central nervous system directly through the olfactory nerves in the olfactory area of the nasal cavity. Early observations of patients with nasal infections revealed that the meninges can also become infected, suggesting there is a direct route to the CNS from the nasal cavity. Researchers have confirmed that transport does indeed occur between the olfactory nerve and the CNS [8, 22]. This suggests that nasally administered drugs can directly target the CNS and potentially avoid both systemic circulation and the blood brain barrier. The mechanisms of drug transport through the olfactory nerve has not been fully elucidated, but there are some reports detailing viruses and heavy metal particles entering the CNS via the olfactory nerve [23, 24].

11.3.3 Characteristics of Nasal Administration

11.3.3.1 Bioavailability

Compared to oral delivery, nasal administration does not need to pass through the gastrointestinal tract to directly reach the site of action. This avoids degradation in the gastrointestinal fluid and the first-pass metabolism from the liver. Since the route of delivery to the CNS is more direct, only a small amount of drug (generally ~a tenth of the oral dose) is needed to reach an effective concentration. For example, intranasal salbutamol can relieve dyspnea in children with bronchospasm. However, the dose for intranasal delivery is only 100 μg compared with 2–4 mg by oral administration. Another example is the antiarrhythmic drug propranolol. Propranolol is greatly affected by first-pass metabolism after oral administration and the bioavailability is only 7–19%. Nasal administration can increase the bioavailability to nearly 100%.

11.3.3.2 Convenience, Compliance, and Costs

Intranasal delivery is typically quite simple. Many patients or their parents can administer medication using this method. Unlike some oral preparations, intranasal methods do not require spacing around meal time. The simplicity and convenience of intranasal delivery allows for higher patient compliance [25]. This treatment method is especially easy for children. Formulations are already prepared to treat the common cold, fever, upper respiratory tract infections, and other common ailments [26, 27]. It is also suitable for some patients who cannot take medications orally for various reasons. Intranasal medication delivery is also quite cost-effective, especially when time and resource use as well as patient satisfaction are concerned [28, 29].

11.3.3.3 Kinetics

Drugs given via nasal administration are absorbed and act quickly. The nasal absorption rate of non-peptide drugs is comparable to that of drugs injected intravenously [30]. The rapid action and convenience of intranasal delivery make it a suitable method for drugs used in emergent situations. For example, nitroglycerin is commonly used for the treatment of angina in patients with coronary heart disease and can relieve pain in 2–5 min and may benefit from intranasal administration [31]. Nasal administration of anticonvulsant drugs such as diazepam and clonazepam can be used in epileptic seizures. Studies comparing the average effective time of diazepam indicated intranasal delivery was significantly faster than intramuscular delivery for the treatment of convulsions in children [32–35].

11.4 Adverse Effects

Adverse effects specific to intranasally delivered medications are infrequent [36]. Some drugs may affect the movement of or may be toxic to cilia within the nose. This may play a role in reducing drug tolerance. The molecular size of intranasally delivered drugs may affect ciliary toxicity [37]. Studies have found that macromolecular drugs are relatively less toxic to nasal cilia, however, certain small molecule synthetic drugs may have a more pronounced effects on nasal ciliary movement [37]. However, drugs with larger molecular weights may not be absorbed as efficiently as drugs with smaller molecular weights [38]. Other drugs may affect the nasal mucosa. The mucosal toxicity of some drugs may limit advances in research and are more appropriately administered by other means. The balance between absorption and mucosal toxicity requires further study.

11.5 Common Uses for Intranasal Medications in Children

Intranasal medications have been used for a variety of purposes including vaccine delivery, rhinosinusitis, seizures, migraines, sedation and analgesia, and delivery of opioid antagonists. In children, the most common use of the intranasal delivery techniques is for sedation and analgesia, anxiolysis, anti-epileptics, and migraine control. A summary of these medications and recommended doses are listed in Table 11.1.

11.5.1 Sedatives and Analgesia

At present, the clinical use of pediatric preoperative medication is usually via intramuscular or intravenous administration. However, the patients are usually awake preoperatively, and fear of injection and the "white coat effect" may cause unwanted changes in blood pressure and heart rate. Amplifying the negative experiences associated with surgery may also have an impact on the patient's psychological development. Intranasal delivery of drugs such as benzodiazepines and opioids might reduce the pre-operative stress.

Midazolam is commonly used for pediatric sedation. The drug can be administered by oral, rectal, intramuscular, intravenous and intranasal routes [54, 55]. Intranasal midazolam is quite useful for procedural sedation. Theroux et al. found that for preschool children requiring laceration repair surgery, 0.4 mg/kg intranasal midazolam could reduce crying and struggle scores compared with intranasal saline placebo or no intervention [56]. Ljungman et al. reported that parents and nurses described less anxiety, discomfort, and procedural problems in children who received intranasal midazolam at 0.2 mg/kg versus placebo. Some of the adverse effects that have been reported for intranasal midazolam include nasal irritation, unpleasant taste, salivation, nausea and vomiting, changes in vision, and gait difficulties [57].

Fentanyl is an ideal intranasal drug because of its high lipophilicity and relatively low molecular weight. Peak plasma concentrations can be reached within 10–15 min after delivery. Borland et al. found that 1.7 mg Hg/kg intranasal fentanyl was equivalent to 0.1 mg/kg intravenous morphine for analgesia in children [39]. Adverse reactions to intranasal fentanyl are rare and include nosebleeds and unpleasant tastes [11, 39]. Other work has also shown that intranasal fentanyl is effective in

11 Intranasal Medication Delivery in Children for Brain Disorders

Table 11.1 Comparative doses by route for intranasal medications

	Intranasal	Intravenous	Oral	Other	References
Sedation and analgesia					
Fentanyl	Dose: 1–2 ug/kg; onset: 5–10 min; duration: Related to blood level; half-life is longer than intravenous delivery respiratory depression last longer than analgesia. Available concentrations: 50 ug/mL; as metered dose for adult patients:100 ug/spray, 400 ug/spray	Dose: 1–4 ug/kg dose (higher doses may be needed in younger children); onset: Immediate; duration: 30 min to 1 h	Approved for patients >16 years	Buccal: approved for patients aged >18 years	[39–41]
Midazolam	Dose: 0.2–0.3 mg/kg (maximum, 10 mg) (use only in patients >6 month of age); onset: 5 min; duration: 30–60 min; available concentrations: 1 mg/mL, 5 mg/mL	Dose: 0.05–0.1 mg/kg (maximum, 10 mg); onset: 1–5 min; duration: 20–30 min	Dose: 0.25–0.5 mg/kg; onset: 10–20 min; duration: Variable		[42–45]
Ketamine	Optimum dose not determined yet in children (off-label); available concentrations: 10, 50, 100 mg/mL	Dose: 0.5–2 mg/kg; onset: 30 s; duration: 5–10 min (recovery 1–2 h)	Dose: 6–10 mg/kg; onset: 1–5 min; duration: 20–30 min		[43, 46, 47]
Anti-epileptics					
Midazolam	Dose: 0.2 mg/kg (maximum,10 mg) (>1 month of age); onset: 5 min; duration: 30–60 min; available concentrations: See above	Dose: Loading 0.06–0.15 mg/kg; infusion: 1–7 ug/kg per min		Buccal dose: 0.25–0.5 mg/kg (maximum, 10 mg) (>3 month of age); onset: Unclear; duration: Unclear	[43, 48, 49]
Lorazepam	(off-label) dose: 0.1 mg/kg (maximum, 4 mg); onset: Unclear; duration: Unclear; available concentrations: 2 and 4 mg/mL	Dose:0.05–0.1 mg/kg (maximum,4 mg/dose); onset: 5 min; duration: 8–12 h			[50]
Opioid antagonist					
Naloxone	No recommendations in children, in adults (off-label); dose: 2 mg onset: 8–13 min; duration: Unclear; available concentrations: 0.4 and 1 mg/mL	Dose: 0.1 mg/kg (maximum, 2 mg); onset: 2 min; duration: 20–60 min			[51, 52]
Antimigraine					
Sumatriptan	Dose: 5, 10, 20 mg (>5 years of age); onset: 15–30 min; duration: Unclear; available concentrations: 5 and 20 mg per 0.1 mL		Not approved for children	Subcutaneous dose: 3–6 mg; (6–18 years of age) onset: 10 min to 2 h; duration: Unclear	[53]

treating pain associated with fractures in children [58, 59]. Another synthetic opioid, sufentanil, has also been administered intranasally for analgesia and sedation in children [60, 61].

Recently, the use of intranasal ketamine in children has received attention [40, 62]. Ketamine is a pediatric analgesic and sedative. It has recently become the focus of research for intranasal administration. Roelofse et al. compared the intranasal administration of 20 Hg sufentanil and 0.3 mg/kg midazolam in healthy children weighing between 15 and 20 kg undergoing dental surgery [63]. They found that the two treatment groups had the same sedative effects.

11.5.2 Seizure Control

Intranasal midazolam also provides an effective treatment option for patients with epilepsy. Midazolam easily crosses the nasal mucosa and blood brain barrier, causing a rapid increase in plasma and cerebrospinal fluid concentrations [64, 65]. Fisgin et al. compared rectal administration with intranasal midazolam and found intranasal midazolam work faster and is more effective at interrupting seizures (60% vs 87%) [48]. Compared to intravenous diazepam, intranasal midazolam has a similar effect (92% vs 88%) and was faster at ceasing seizure activity [66, 67]. In addition, the use of intranasal midazolam and lorazepam is safe for the treatment of seizures for use by patients. Ahmad et al. compared intranasal lorazepam and intramuscular injection of lorazepam in 160 pediatric patients in rural Africa, most of whom had long-term seizures due to cerebral malaria or bacterial meningitis [68]. Intranasal lorazepam stopped 75% of seizures within a few minutes, while intramuscular paraldehyde was effective only 61% of the time. Holsti et al. compared the treatment of seizures with rectal diazepam or intranasal midazolam in children [49]. Pre-hospital seizure control rate (62% vs 28%), emergency intubation rate (11% vs 42%), admission requirements (40% vs 89%) and ICU admission rate (16% vs 59%). Compared to rectal administration of diazepam, the intranasal midazolam group had significantly better outcomes [32, 34]. Family epilepsy treatment by parents at home is also effective, and safer than rectal diazepam [69, 70]. Cumulatively, these findings suggest that intranasal midazolam is a favorable treatment option for epilepsy in children.

11.5.3 Opioid Antagonists

Rapid administration of naloxone can alleviate the symptoms of respiratory depression caused by opioid overdose. Traditionally, naloxone is administered intravenously [71, 72]. Excess opioid use in the pediatric population is usually due to accidental intake. However, since peripheral venous access is often difficult to obtain in people who abuse opioids, intranasal administration of naloxone may be a simple and rapid alternative. Intranasal delivery of naloxone provides a similar bioavailability and onset time [73]. In addition, compared to intravenous administration, intranasal administration can reduce the potential for needle stick injuries [74]. Because patients with a history of intravenous drug abuse tend to have a higher risk of infectious disease, this situation requires the protection of medical personnel from puncture injuries [74]. In adult prehospital patients, Barton et al. found no difference in time to onset between intranasal and intravenous naloxone administration by paramedics [75]. Robertson et al. found that although the clinical response of intravenous naloxone was faster than that of intranasal, there is no significant difference in the time from initial contact to clinical response to cessation of clinical response [76]. This may be due to the time needed to establish venous access [77].

11.5.4 Anti-Migraine

Migraine is a common chronic and recurrent headache disorder in the pediatric population. The age of onset is commonly 6–12 years old. The incidence in males is slightly more than that in females before the age of 10, however, the rate in adolescent females is higher than that in males. Oral medication is usually administered as a first line of therapy. When oral treatment fails, intranasal drugs can be used instead of intravenous therapy in certain situations. The most common intranasal anti-migraine drug is sumatriptan [78]. A comparison of intranasal sumatriptan at 5, 10, and 20 mg with normal saline placebo found that sumatriptan offered more relief than placebo at 1 h for those receiving 10 and 20 mg and more relief than placebo at 2 h for those receiving 5 mg [79]. Ahonen and Lewis found that intranasal sumatriptan was more effective in relieving migraine than placebo [80, 81]. The side effects of intranasal triptans include unpleasant taste, nasal discomfort, and congestion [82]. The use of intranasal lidocaine for the treatment of acute migraine has not been fully studied in children. Some studies in adults have shown that lidocaine can be used to stop migraine, however, the data is limited [83, 84].

11.6 Conclusions

Intranasal delivery offers an attractive alternative to invasive drug delivery for delivering analgesia, anxiolytics, and anticonvulsants to pediatric patients. The major advantages of intranasal delivery include the straightforward and needle-free application modality and the permeable application site in the nasal cavity that allows for a rapid onset of local and systemic drug activity. Furthermore, intranasal delivery may reduce medical staff resource use, eliminate needle-stick exposure risk, and lead to improved patient and parent satisfaction. Pediatricians, pediatric emergency physicians, and emergency medical services should consider adopting this delivery method for medications and indications that are appropriate to their practice setting.

References

1. Guennoun R, et al. Intranasal administration of progesterone: a potential efficient route of delivery for cerebroprotection after acute brain injuries. Neuropharmacology. 2018;145(Pt B):283–91.
2. Wolfe TR, Braude DA. Intranasal medication delivery for children: a brief review and update. Pediatrics. 2010;126(3):532–7.
3. Talon MD, et al. Intranasal dexmedetomidine premedication is comparable with midazolam in burn children undergoing reconstructive surgery. J Burn Care Res. 2009;30(4):599–605.
4. Corrigan M, Wilson SS, Hampton J. Safety and efficacy of intranasally administered medications in the emergency department and prehospital settings. Am J Health Syst Pharm. 2015;72(18):1544–54.
5. Bitter C, Suter-Zimmermann K, Surber C. Nasal drug delivery in humans. Curr Probl Dermatol. 2011;40:20–35.
6. Barash PG, et al. Is cocaine a sympathetic stimulant during general anesthesia? JAMA. 1980;243(14):1437–9.
7. Fantacci C, et al. Intranasal drug administration for procedural sedation in children admitted to pediatric emergency room. Eur Rev Med Pharmacol Sci. 2018;22(1):217–22.
8. Kanazawa T. [Development of noninvasive drug delivery systems to the brain for the treatment of brain/central nervous system diseases]. Yakugaku Zasshi. 2018;138(4):443–50.
9. Chamanza R, Wright JA. A review of the comparative anatomy, histology, physiology and pathology of the nasal cavity of rats, mice, dogs and non-human primates. Relevance to inhalation toxicology and human health risk assessment. J Comp Pathol. 2015;153(4):287–314.
10. Stenner M, Rudack C. Diseases of the nose and paranasal sinuses in child. GMS Curr Top Otorhinolaryngol Head Neck Surg. 2014;13:Doc10.
11. Grassin-Delyle S, et al. Intranasal drug delivery: an efficient and non-invasive route for systemic administration: focus on opioids. Pharmacol Ther. 2012;134(3):366–79.
12. Koskenkorva T, Kristo A. [It's normal—structural and functional variations of nose and paranasal sinuses]. Duodecim. 2012;128(2):225–9.
13. Ooi EH, Wormald PJ, Tan LW. Innate immunity in the paranasal sinuses: a review of nasal host defenses. Am J Rhinol. 2008;22(1):13–9.
14. Imamura F, Hasegawa-Ishii S. Environmental toxicants-induced immune responses in the olfactory mucosa. Front Immunol. 2016;7:475.
15. Yeh CY, et al. Activated human nasal epithelial cells modulate specific antibody response against bacterial or viral antigens. PLoS One. 2013;8(2):e55472.
16. Petrov VV, Tepliy DL. [The functional state of nasal cavity in the aspect of structural-functional changes of the human organism in postnatal ontogenesis.]. Adv Gerontol. 2017;30(5):739–44.
17. Dahl R, Mygind N. Anatomy, physiology and function of the nasal cavities in health and disease. Adv Drug Deliv Rev. 1998;29(1–2):3–12.
18. Pires A, et al. Intranasal drug delivery: how, why and what for? J Pharm Pharm Sci. 2009;12(3):288–311.
19. Snidvongs K, Thanaviratananich S. Update on intranasal medications in rhinosinusitis. Curr Allergy Asthma Rep. 2017;17(7):47.
20. Phukan K, et al. Nanosized drug delivery systems for direct nose to brain targeting: a review. Recent Pat Drug Deliv Formul. 2016;10(2):156–64.
21. Mistry A, Stolnik S, Illum L. Nanoparticles for direct nose-to-brain delivery of drugs. Int J Pharm. 2009;379(1):146–57.
22. Crowe TP, et al. Mechanism of intranasal drug delivery directly to the brain. Life Sci. 2018;195:44–52.
23. Marianecci C, et al. Drug delivery in overcoming the blood-brain barrier: role of nasal mucosal grafting. Drug Des Devel Ther. 2017;11:325–35.
24. Khan AR, et al. Progress in brain targeting drug delivery system by nasal route. J Control Release. 2017;268:364–89.

25. Freitag FG, Shumate DA. The efficacy and safety of sumatriptan intranasal powder in adults with acute migraine. Expert Rev Neurother. 2016;16(7):743–7.
26. Au CC, Branco RG, Tasker RC. Management protocols for status epilepticus in the pediatric emergency room: systematic review article. J Pediatr. 2017;93(Suppl 1):84–94.
27. DeMayo MM, et al. A review of the safety, efficacy and mechanisms of delivery of nasal oxytocin in children: therapeutic potential for autism and Prader-Willi syndrome, and recommendations for future research. Paediatr Drugs. 2017;19(5):391–410.
28. Borland ML, Clark LJ, Esson A. Comparative review of the clinical use of intranasal fentanyl versus morphine in a paediatric emergency department. Emerg Med Australas. 2008;20(6):515–20.
29. Young VN, Smith LJ, Rosen CA. Comparison of tolerance and cost-effectiveness of two nasal anesthesia techniques for transnasal flexible laryngoscopy. Otolaryngol Head Neck Surg. 2014;150(4):582–6.
30. Parvizrad R, et al. Comparing the analgesic effect of intranasal with intravenous ketamine in isolated orthopedic trauma: a randomized clinical trial. Turk J Emerg Med. 2017;17(3):99–103.
31. Luthringer R, et al. Rapid absorption of sumatriptan powder and effects on glyceryl trinitrate model of headache following intranasal delivery using a novel bi-directional device. J Pharm Pharmacol. 2009;61(9):1219–28.
32. Charalambous M, et al. Intranasal midazolam versus rectal diazepam for the management of canine status epilepticus: a multicenter randomized parallel-group clinical trial. J Vet Intern Med. 2017;31(4):1149–58.
33. Maglalang PD, et al. Rescue therapies for seizure emergencies: new modes of administration. Epilepsia. 2018;59:207–15.
34. Nunley S, et al. Healthcare utilization characteristics for intranasal midazolam versus rectal diazepam. J Child Neurol. 2018;33(2):158–63.
35. Zelcer M, Goldman RD. Intranasal midazolam for seizure cessation in the community setting. Can Fam Physician. 2016;62(7):559–61.
36. Campbell C, et al. Drug development of intranasally delivered peptides. Ther Deliv. 2012;3(4):557–68.
37. Fortuna A, et al. Intranasal delivery of systemic-acting drugs: small-molecules and biomacromolecules. Eur J Pharm Biopharm. 2014;88(1):8–27.
38. Al Bakri W, et al. Overview of intranasally delivered peptides: key considerations for pharmaceutical development. Expert Opin Drug Deliv. 2018;15(10):991–1005.
39. Borland M, et al. A randomized controlled trial comparing intranasal fentanyl to intravenous morphine for managing acute pain in children in the emergency department. Ann Emerg Med. 2007;49(3):335–40.
40. Miller JL, et al. Sedation and analgesia using medications delivered via the extravascular route in children undergoing laceration repair. J Pediatr Pharmacol Ther. 2018;23(2):72–83.
41. Adelgais KM, et al. Intranasal fentanyl and quality of pediatric acute care. J Emerg Med. 2017;53(5):607–615 e2.
42. Baldwa NM, et al. Atomised intranasal midazolam spray as premedication in pediatric patients: comparison between two doses of 0.2 and 0.3 mg/kg. J Anesth. 2012;26(3):346–50.
43. Hosseini Jahromi SA, et al. Comparison of the effects of intranasal midazolam versus different doses of intranasal ketamine on reducing preoperative pediatric anxiety: a prospective randomized clinical trial. J Anesth. 2012;26(6):878–82.
44. Mellion SA, et al. Evaluating clinical effectiveness and pharmacokinetic profile of atomized intranasal midazolam in children undergoing laceration repair. J Emerg Med. 2017;53(3):397–404.
45. Sulton C, et al. The use of intranasal dexmedetomidine and midazolam for sedated magnetic resonance imaging in children: a report from the pediatric sedation research consortium. Pediatr Emerg Care. 2017. https://doi.org/10.1097/PEC.0000000000001199.
46. Tsze DS, et al. Intranasal ketamine for procedural sedation in pediatric laceration repair: a preliminary report. Pediatr Emerg Care. 2012;28(8):767–70.

47. Bahetwar SK, et al. A comparative evaluation of intranasal midazolam, ketamine and their combination for sedation of young uncooperative pediatric dental patients: a triple blind randomized crossover trial. J Clin Pediatr Dent. 2011;35(4):415–20.
48. Fisgin T, et al. Effects of intranasal midazolam and rectal diazepam on acute convulsions in children: prospective randomized study. J Child Neurol. 2002;17(2):123–6.
49. Holsti M, et al. Prehospital intranasal midazolam for the treatment of pediatric seizures. Pediatr Emerg Care. 2007;23(3):148–53.
50. Arya R, et al. Intranasal versus intravenous lorazepam for control of acute seizures in children: a randomized open-label study. Epilepsia. 2011;52(4):788–93.
51. Merlin MA, et al. Intranasal naloxone delivery is an alternative to intravenous naloxone for opioid overdoses. Am J Emerg Med. 2010;28(3):296–303.
52. Vanky E, et al. Pharmacokinetics after a single dose of naloxone administered as a nasal spray in healthy volunteers. Acta Anaesthesiol Scand. 2017;61(6):636–40.
53. Winner P, et al. A randomized, double-blind, placebo-controlled study of sumatriptan nasal spray in the treatment of acute migraine in adolescents. Pediatrics. 2000;106(5):989–97.
54. Ng E, Taddio A, Ohlsson A. Intravenous midazolam infusion for sedation of infants in the neonatal intensive care unit. Cochrane Database Syst Rev. 2017;1:CD002052.
55. Thomas A, et al. Non-intravenous sedatives and analgesics for procedural sedation for imaging procedures in pediatric patients. J Pediatr Pharmacol Ther. 2015;20(6):418–30.
56. Theroux MC, et al. Efficacy of intranasal midazolam in facilitating suturing of lacerations in preschool children in the emergency department. Pediatrics. 1993;91(3):624–7.
57. Ljungman G, et al. Midazolam nasal spray reduces procedural anxiety in children. Pediatrics. 2000;105(1 Pt 1):73–8.
58. Saunders M, Adelgais K, Nelson D. Use of intranasal fentanyl for the relief of pediatric orthopedic trauma pain. Acad Emerg Med. 2010;17(11):1155–61.
59. Furyk JS, Grabowski WJ, Black LH. Nebulized fentanyl versus intravenous morphine in children with suspected limb fractures in the emergency department: a randomized controlled trial. Emerg Med Australas. 2009;21(3):203–9.
60. Nielsen BN, et al. Intranasal sufentanil/ketamine analgesia in children. Paediatr Anaesth. 2014;24(2):170–80.
61. Hitt JM, et al. An evaluation of intranasal sufentanil and dexmedetomidine for pediatric dental sedation. Pharmaceutics. 2014;6(1):175–84.
62. AlSarheed MA. Intranasal sedatives in pediatric dentistry. Saudi Med J. 2016;37(9):948–56.
63. Roelofse JA, et al. Intranasal sufentanil/midazolam versus ketamine/midazolam for analgesia/sedation in the pediatric population prior to undergoing multiple dental extractions under general anesthesia: a prospective, double-blind, randomized comparison. Anesth Prog. 2004;51(4):114–21.
64. Henry RJ, et al. A pharmacokinetic study of midazolam in dogs: nasal drop vs. atomizer administration. Pediatr Dent. 1998;20(5):321–6.
65. Malinovsky JM, et al. Plasma concentrations of midazolam after i.v., nasal or rectal administration in children. Br J Anaesth. 1993;70(6):617–20.
66. Lahat E, et al. Comparison of intranasal midazolam with intravenous diazepam for treating febrile seizures in children: prospective randomised study. BMJ. 2000;321(7253):83–6.
67. Mahmoudian T, Zadeh MM. Comparison of intranasal midazolam with intravenous diazepam for treating acute seizures in children. Epilepsy Behav. 2004;5(2):253–5.
68. Ahmad S, et al. Efficacy and safety of intranasal lorazepam versus intramuscular paraldehyde for protracted convulsions in children: an open randomised trial. Lancet. 2006;367(9522):1591–7.
69. Wilson MT, Macleod S, O'Regan ME. Nasal/buccal midazolam use in the community. Arch Dis Child. 2004;89(1):50–1.
70. Harbord MG, et al. Use of intranasal midazolam to treat acute seizures in paediatric community settings. J Paediatr Child Health. 2004;40(9–10):556–8.
71. Weiner SG, et al. Use of intranasal naloxone by basic life support providers. Prehosp Emerg Care. 2017;21(3):322–6.

72. Warrington SE, Kuhn RJ. Use of intranasal medications in pediatric patients. Orthopedics. 2011;34(6):456.
73. Costantino HR, et al. Intranasal delivery: physicochemical and therapeutic aspects. Int J Pharm. 2007;337(1–2):1–24.
74. Centers for Disease Control and Prevention (CDC). Integrated prevention services for HIV infection, viral hepatitis, sexually transmitted diseases, and tuberculosis for persons who use drugs illicitly: summary guidance from CDC and the U.S. Department of Health and Human Services. MMWR Recomm Rep. 2012;61(RR-5):1–40.
75. Barton ED, et al. Efficacy of intranasal naloxone as a needleless alternative for treatment of opioid overdose in the prehospital setting. J Emerg Med. 2005;29(3):265–71.
76. Robertson TM, et al. Intranasal naloxone is a viable alternative to intravenous naloxone for prehospital narcotic overdose. Prehosp Emerg Care. 2009;13(4):512–5.
77. Bailey AM, et al. Review of intranasally administered medications for use in the emergency department. J Emerg Med. 2017;53(1):38–48.
78. Miyake MM, Bleier BS. The blood-brain barrier and nasal drug delivery to the central nervous system. Am J Rhinol Allergy. 2015;29(2):124–7.
79. Lipton RB, et al. DFN-02 (sumatriptan 10 mg with a permeation enhancer) nasal spray vs placebo in the acute treatment of migraine: a double-blind, placebo-controlled study. Headache. 2018;58(5):676–87.
80. Ahonen K, et al. Nasal sumatriptan is effective in treatment of migraine attacks in children: a randomized trial. Neurology. 2004;62(6):883–7.
81. Lewis DW, et al. Efficacy of zolmitriptan nasal spray in adolescent migraine. Pediatrics. 2007;120(2):390–6.
82. Priprem A, et al. Intranasal melatonin nanoniosomes: pharmacokinetic, pharmacodynamics and toxicity studies. Ther Deliv. 2017;8(6):373–90.
83. Maizels M, Geiger AM. Intranasal lidocaine for migraine: a randomized trial and open-label follow-up. Headache. 1999;39(8):543–51.
84. Maizels M, et al. Intranasal lidocaine for treatment of migraine: a randomized, double-blind, controlled trial. JAMA. 1996;276(4):319–21.

Index

A
Acetylcholinesterase inhibitors (AChEIs), 118
Age-related macular degeneration (AMD), 13, 23
Albumin, 30
Allopregnanolone (Allo), 129
Alzheimer's disease (AD)
 angiotensin receptor blocker, 126
 cell-based therapy, 129, 130
 chelators, 118
 curcumin, 125
 DFO, 125
 galantamine, 122
 hormone, 128, 129
 Hup A, 123
 immunization, 129
 insulin, 124
 neurodegenerative disorder, 118
 neurotrophic factors, 126, 127
 peptide, 128
 personality and behavior change, 118
 physostigmine, 123
 PIP, 126
 quercetin, 124
 R-flurbiprofen, 125
 rivastigmine, 122
 tacrine, 121
 TFB, 123
Aneurysmal subarachnoid hemorrhage (aSAH)
 cerebral ischemia, 62
 delayed ischemic injury, 57
 intranasal stem cell treatment, 60
 nimodipine-loaded lipopluronics micelles, 60, 61
Angiogenesis, 10, 12, 15, 22–24, 92
Angiotensin-converting enzyme (ACE), 22, 23
Angiotensin receptor blocker, 126
Animal model, 78
 ICH, 44, 50, 52
 ischemic stroke, 79–80
Apelin, 10
Apo-transferrin (aTf), 68
Arteriogenesis, 10, 15, 22, 23, 25
Aryl hydrocarbon receptor (AhR), 14
Axonal remodeling, 108, 109

B
Basic fibroblast growth factor (bFGF), 82, 127
B-cell lymphoma 2 (Bcl-2), 11
Bioactive factors (BFs), 36
Blood-brain barrier (BBB), 28–30, 33, 36, 44, 45, 47, 48, 50, 57, 58, 66, 70, 92, 105, 118
Bone marrow-derived MSCs (BMSCs), 34, 80, 81
Brain-derived neurotrophic factor (BDNF), 34, 108
Brain injury
 history and development, 136
 intraosseous delivery, 136
 liver, 136
 needle-stick injury, 136
Brainstem, 76, 77

C

Calpain, 28
cAMP response element-binding protein
    (CREB), 14, 24
Carbon monoxide, 10
Cardiac arrest, 3, 6
CC chemokine receptor-7 (CCR-7), 11
Cell-based therapy, 129, 130
Centers for Disease Control (CDC), 28
Central nervous system (CNS), 66, 92, 138
Cerebral ischemia, 78
Cerebral ischemic stroke, 68, 71
Cerebrospinal fluid (CSF), 66, 67
Chitosan nanoparticles (CS-NPs), 126
Choline acetyltransferase activity, 127
Chondroitin sulfate proteoglycans
    (CSPGs), 108
Chondroitinase ABC (ChABC), 109
Clonazepam, 139
Clustered regularly interspaced short
    palindromic repeats (CRISPR), 25
Cobalt protoporphyrin, 10, 14
Conditioning, 9, 14
CoolStat Device, 4–6
Corticospinal tract (CST), 106–108
Curcumin, 125
C-X-C chemokine receptor type 4 (CXCR-4),
    11, 15, 22–24
Cyclodextrins, 30
Cytokines, 94

D

Deferoxamine (DFO), 50, 51, 125
1-Desamino-8-D-arginine vasopressin
    (DDAVP), 35
Diazepam, 139
Diazoxide, 10
Double strand break (DSB), 25
Drug administration, 46, 47
Dry air, 2, 4, 5

E

Early brain injury, 58, 62
Electromyograms (EMG), 106
Embryonic stem cells (ESCs), 10–13, 15
Endonucleases, 28
Endothelial cells, 102
Endothelial progenitor cells (EPCs), 10, 15, 22
Endothelial/extracellular matrix (ECM), 14, 15
Epithelial barriers, 76, 77
Erythropoietin (EPO), 10–12, 15, 22–24, 32,
    61, 62, 65, 68, 71

Excitotoxicity, 45
Exogenous recombinant human erythropoietin
    (rHu-Epo), 78
Experimental ICH, 45, 46, 48
Extracellular signal-regulated kinase (ERK),
    22, 109

F

Fentanyl, 140

G

Galantamine, 122
Gamma aminobutyric acid (GABA), 13
Gelatin nanoparticles (GNPs), 82
Glucagon-like peptide-1 (GLP-1), 78, 124
Granulocyte colony stimulating factor
    (G-CSF), 15, 78

H

Heat shock protein (Hsp), 10
Hematopoietic stem cells (HSCs), 10, 22
Heme oxygenase (HO), 45
Hemoglobin, 45
High-mobility-group-box-1 (HMGB1), 78,
    81, 95
Hippocampus, 92
Hormone, 128, 129
Horseradish peroxidase (HRP), 77
Human acidic fibroblast growth factor
    (haFGF), 127
Human recombinant erythropoietin (rHu-
    EPO), 32
Huperizin A (Hup A), 123
Hydrogen dioxide (H_2O_2), 10
Hydrogen sulfide (H_2S), 10
Hypothermia, *see* Therapeutic hypothermia
Hypoxia-primed stem cell
    after fetal tracheal occlusion, 11
    conditioning medicine and cell survival
        mechanisms, 13, 14
    ESCs, 10
    genetic engineering techniques, 9
    hypoxic and ischemic models, 9
    inflammation and immune responses,
        24, 25
    intranasal cell therapy, 12, 13
    ischemic/hypoxic preconditioning, 10
    low-oxygen sensor, 11
    mitochondrial mechanisms, 14
    and neuroplasticity, 23, 24
    pathophysiology, 11

Index 151

post-ischemic flow recovery, 15, 22
progenitor cell-based therapies, 10
PSCs, 10
regeneration, 24, 25
stem cell research, 9
stroke, 10
transplantation, 10

I

Immunization, 129
Immunoglobulins, 137
Inducible nitric oxide synthase (iNOS), 33
Insulin, 124
Insulin-like growth factor-1 (IGF-1), 10, 22, 31, 32, 62, 68
Interleukin-17A (IL-17A), 92
  EPO, cerebral ischemia, 96
  growth factor, 93
  intranasal application, 96
  ischemic stroke, 95, 96
  neurogenesis and functional recovery, 94
  neurologic disorders, 92, 93
Intracerebral hemorrhage (ICH)
  animal models, 52
  DFO, 50, 51
  drug administration, 46, 47
  excitotoxicity, 45
  hemoglobin and iron, 45, 46
  intranasal therapeutics, 47
  ischemic stroke, 44
  mortality rate, 44
  neuroprotective considerations, 46
  pathophysiology, 44
  recombinant proteins, 48
  risk factors, 43
  secondary injury, 51
  small molecules, 48, 49
  stem cells, 49
  thrombin, 45
Intracerebral hemorrhagic (ICH) stroke, 35
Intra-cerebroventricular (ICV), 59, 68, 92
Intracortical microstimulation (ICMS), 106
Intranasal drug delivery, 47, 52
  administration, 60
  advantages and challenges, 118
  BBB, 29, 30
  BCSFB, 29
  clinical trials, 35, 36
  CNS and bypasses, 28
  CNS disorders, 36
  α-cyclodextrin, 31

  β-cyclodextrin, 31
  disadvantages, 36
  erythropoietin, 32
  factors and reagents, 34, 35
  IGF-1, 31, 32
  intracerebral hemorrhagic stroke, 35
  MSCs, 34
  nasal cavity and olfactory tissue, 30
  neurologic damage after stroke, 28, 29
  non-target regions, 31
  osteopontin, 33
  preclinical translational stroke research, 36
  stroke cases, 27
  TGF, 33, 34
  TIA, 28
  tPA, 28
  transport, nose to brain, 119
Intranasal medication delivery
  adverse effects, 139
  anti-migraine, 143
  nasal administration
    bioavailability, 138
    convenience, compliance and costs, 139
    kinetics, 139
  nasal mucosa, 137
  opioid antagonists, 142, 143
  sedatives and analgesia, 140, 142
  seizure control, 142
Ischemic stroke, 10, 15, 22–24
  clinical diagnosis, 28
  IN treatment
    erythropoietin, 32
    factors and reagents, 34
    IGF-1, 31, 32
    MSCs, 34
    osteopontin, 33
    TGF, 33, 34
  intranasal administration
    gene vectors, 81
    proteins and peptides, 78
    small molecules, 79–82
    stem cell, 80, 81

K

Kruppel-like factor 4 (KLF-4), 12

L

Lamina propria, 76, 77, 85
Lipopluronics micelles, 60, 61
Lipoprotein receptor-related protein 1 (LRP1), 109

M

Manganese superoxide dismutase (MnSOD), 11
Matrix metalloproteinase (MMP), 95
Matrix metalloproteinases-9 (MMP-9), 108
Mechanistic target of rapamycin (mTOR), 14
Mesenchymal stem/stromal cells (MSCs), 10, 12–15, 22–25, 34, 81
Microemulsion (ME), 122
Microthrombosis, 57
Midazolam, 140, 142
Middle cerebral artery occlusion (MCAO), 31–35, 78, 81, 82, 93, 102
Modified neurological severity score (mNSS) test, 105
Morris water maze (MWM) test, 105
MSC over-expressing brain-derived neurotrophic factor (MSC-BDNF), 81
Mucoadhesive microemulsions (MMEs), 122
Mucociliary clearance (MCC), 84

N

N-acetylglucosamine, 127
Naloxone, 142
Nasal cavity, 93, 138
Netrin-1 (NTN-1), 60
Neural progenitor cell (NPC), 95
Neurocritical care (NCCU), 3
Neuro-erythropoietin (Neuro-EPO), 78
Neurogenesis, 10, 15, 22, 23
Neuroinflammatory diseases, 96
Neurologic disorders, 92
Neuroprotection, 6, 11, 15, 23, 61
Neurotrophic factors, 126, 127
Neurotrophin, 48, 49
Nimodipine, 60, 61
Nitric oxide (NO), 95
N-methyl-D-aspartate (NMDA) receptor, 95, 109
Nuclear factor kB (NF-kB), 118

O

Olfactory bulbs, 77
Olfactory sensory neuron (OSN), 76, 77
Oligodendrocyte progenitor cells (OPCs), 10, 24
Opioids, 81, 82
Osteopontin (OPN), 33, 59, 82
Oxygen glucose deprivation (OGD) injury, 103

P

PEGylation, 70
Peptide, 128
Perflurocarbon (PFC), 3, 4
Perihematomal edema, 44, 45
Peroxisome proliferator-activated receptor-g (PPAR-g), 118
Physostigmine, 123
Piperine (PIP), 126
Pluripotent stem cells (PSCs), 10, 12
Polyethylene glycol (PEG), 33, 70
Preconditioning, 9–11, 15, 22–25
Primarily plasminogen activator inhibitor-1 (PAI-1), 103
Protease-activated receptors (PARs), 45
Pseudorabies virus (PRV), 107

Q

Quercetin, 124
Quinpirole, 35

R

Reactive oxygen species (ROS), 12–14, 28, 29
Recombinant human erythropoietin (rHu-EPO), 93
Recombinant human tissue plasminogen activator (rh-tPA), 104
Recombinant netrin-1, 60
Recombinant Osteopontin (rOPN) SAH, 59
Red fluorescent protein (RFP), 107
Rete mirabile, 4
Reticulo-endothelial system, 61
R-flurbiprofen, 125
Rhinochill device, 3
   acute brain injury, 3
   biologic effects, 4
   in cardiac arrest, 3
   nasopharyngeal cooling, 4
   neurocritical care setting, 3
   PFC chemicals, 3
Rivastigmine, 122
Rostral migratory stream (RMS), 77

S

Single-chain variable fragment (scFv), 129
Small molecules, 48, 49
Solanum tuberosum lectin (STL), 127
Stem cell, 49, 80, 81
Strokes
   hypoxia-primed stem cell transplantation, 10

immunomodulatory and anti-inflammatory effects, 13
intranasal administration
  anatomy difference, rodents and human, 84
  factors affecting delivery, 83, 84
  issues, 83
  local side effect, 84
  merits, 82, 83
  microsphere and nanoemulsions, 85
  muco-adhesive agents, 85
  NeuroEPo, 85
peptides, 68, 69
peri-infarct regions, 22
psychiatric disorders, 66
regenerative neuronal circuits, 23
stem cell transplantation, 24
treatment, 14, 68, 70
VEGF and EPO, 15
Stromal-derived factor-1 (SDF-1), 10, 15, 22, 23
Subarachnoid hemorrhage (SAH)
  and astrocyte, 60
  EPO, 61, 62
  experimental, 60
  MSCs, 60
  OPN, 59
  rOPN, 59
  VitD3, 59
Subgranular zone (SGZ), 23
Subventricular zone (SVZ), 92
Synaptogenesis, 10, 22, 24

T
Tacrine, 121
Tarenflurbil (TFB), 123
Tat-NEMO-binding domain (Tat-NBD), 68
TGF-β/SMAD-3 signaling, 22
Therapeutic hypothermia
  anterior and posterior ethmoid arteries, 1
  brain temperature, 2
  Camino catheter, 2
  cranial sinuses, 2
  mammals without carotid rate, 2
  nasal mucosa, 1
  nasal turbinates, 2
  nasopharyngeal space, 1
  nasopharynx, 2
  Rhinochill device (*see* Rhinochill device)
  transnasal dry air (*see* Transnasal dry air)
  turbinates, 1
Therapeutic peptides
  intranasal delivery
    administration regimens, 70

  advantages, 67
  mechanisms, 66, 67
  non-invasive approach, 71
Thrombin, 45
Tissue plasminogen activator (tPA), 28, 34, 68, 71, 92
  axonal remodeling, 108, 109
  CST, 106–108
  endothelial cells, 102
  hippocampal neurons, 102, 103
  hippocampal-related behavioral tasks, 102
  intranasal administration, 105
  intranasal delivery, brain, 103, 104
  intravenous administration, 102
  ischemia-induced apoptosis, 102
  ischemic stroke, 102, 103
  neurite growth, 103
  neuronal pathologies, 102
  neurons and glial cells, 102
  neuroserpin, 102
  non-thrombotic models, 103
  rtPA treatment, 103
Transcription activator-like effector nucleases (TALENs), 25
Transcytosis, 119
Transforming growth factor (TGF), 33, 34
Transforming growth factor-1 (TGF-1), 12
Transient ischemic attack (TIA), 28, 36
Transnasal dry air
  humidification, 4
  nasal mucus, 4
  peri-operative setting, 4
  PFC based cooling, 6
  porcine brain, 5
  rete mirabile, 4
Transplantation, 10, 12–14, 23–25
Transport
  CNS, 77, 78
  nasal lamina propria, brain entry, 77
  olfactory/respiratory epithelial barriers, 76, 77
Traumatic brain injury (TBI), 103

V
Vascular endothelial growth factor (VEGF), 11, 12, 15, 22, 24
Vasoactive intestinal peptide (VIP), 128
Vasospasm, 57–61
Ventricle subventricular zone (SVZ), 23, 24

X
Xingnaojing microemulsion, 35

Printed by Printforce, the Netherlands